T0351021

Autonomous Positioning of Piezoactuated Mechanism for Biological Cell Puncture

Autonomous Positioning of Piezoactuated Mechanism for Biological Cell Puncture gives a systematic and almost self-contained description of the many facets of advanced design, optimization, modeling, system identification, and advanced control techniques for positioning of the cell puncture mechanism with a piezoelectric actuator in micro/nanorobotics systems.

To achieve biomedical applications, reliability design, modeling, and precision control are essential for developing engineering systems. With the advances in mechanical design, dynamic modeling, system identification, and control techniques, it is possible to expand the advancements in reliability design, precision control, and quick actuation of micro/nanomanipulation systems to the robot's applications at the micro- and nanoscales, especially for biomedical applications.

This book unifies existing and emerging techniques concerning advanced design, modeling, and advanced control methodologies in micropuncture of biological cells using piezoelectric actuators with their practical biomedical applications.

The book is an essential resource for researchers within robotics, mechatronics, biomedical engineering, and automatic control society, including both academic and industrial parts.

KEY FEATURES

- Provides a series of the latest results in, including but not limited to, design, modeling, and control of micro/nanomanipulation systems utilizing piezoelectric actuators
- Gives recent advances of theory, technological aspects, and applications of advanced modeling, control, and actuation methodologies in cell engineering applications
- Presents simulation and experimental results to reflect the micro/nanomanipulation practice and validate the performances of the developed design, analysis, and synthesis approaches

Autonomous Systems and Applications

Series Editor- Hamid Reza Karimi

Autonomous Positioning of Piezoactuated Mechanism for Biological Cell Puncture
Mingyang Xie and Shengdong Yu

Autonomous Positioning of Piezoactuated Mechanism for Biological Cell Puncture

Mingyang Xie

Shengdong Yu

CRC Press
Taylor & Francis Group
Boca Raton London New York

CRC Press is an imprint of the
Taylor & Francis Group, an **informa** business

First edition published 2023
by CRC Press
6000 Broken Sound Parkway NW, Suite 300, Boca Raton, FL 33487-2742

and by CRC Press
4 Park Square, Milton Park, Abingdon, Oxon, OX14 4RN

CRC Press is an imprint of Taylor & Francis Group, LLC

ISBN: 978-1-032-27720-2 (hbk)
ISBN: 978-1-032-27777-6 (pbk)
ISBN: 978-1-003-29403-0 (ebk)

DOI: 10.1201/9781003294030

Typeset in Times
by SPi Technologies India Pvt Ltd (Straive)

Contents

Preface

Cell micromanipulation is a kind of microbiological operation technology that can complete the gene operation of cells or embryos by using a biological microscope. This process involves various cell operations, such as localization, nuclear transfer, cutting, and injection. As the most widely used technology in cell micromanipulation, cell micropuncture has an irreplaceable medical value in the field of physiology, pathology, and pharmacology. Therefore, this technology is widely favored by medical scientists. Biological cell is the basic structural unit of life science and is very fragile with deformable membrane. Therefore, the operation instrument must have high precision to realize cell micropuncture. Moreover, due to the changes of physical characteristics during the cell development process, the cell micromanipulation technology must automatically adapt to the application of such characteristics. Furthermore, the micro-force during the procedure of cell puncture must be well controlled for the assurance of biological cell survival.

Considering the technical requirements and complex application environment of the cell micromanipulation technology, this manuscript first introduces the principle of piezoelectric actuator, which is followed by the design and optimization of piezo-actuated cell puncture mechanism. Then, the dynamics of the developed cell puncture mechanism is identified and modeled. Furthermore, the implementations of high performance position and hybrid force-position controllers based on the dynamic model of the cell puncture mechanism are formulated. Finally, biological cell puncture experiments demonstrate the effectiveness of the proposed mechanical design, optimization, modeling, and hybrid force-position control techniques for positioning of cell puncture mechanism with piezoelectric actuator.

MATLAB® is a registered trademark of The Math Works, Inc. For product information, please contact:

The Math Works, Inc.
3 Apple Hill Drive
Natick, MA 01760-2098
Tel: 508-647-7000
Fax: 508-647-7001
E-mail: info@mathworks.com
Web: http://www.mathworks.com

Authors

Dr. Mingyang Xie, Nanjing University of Aeronautics and Astronautics, College of Automation Engineering.

Prof. Xie is a scholar in the field of micro/nanorobot and micro/nanomanipulation. He has published 1 academic book, 30+ academic papers, including 10+ papers in the top journals and conferences in the field of robotics, such as, *IEEE/ASME Transactions on Mechatronics, IEEE Transactions on Automation Science and Engineering, IEEE Transactions on Control Systems Technology, IEEE Transactions on Biomedical Engineering*, ICRA, IROS, and AIM.

Prof. Xie is currently the associate editor of the IET Electronics Letters, Designs, and guest editor of the *International Journal of Advanced Manufacturing Technology*. He also serves as the workshop organizer of IEEE/RSJ International Conference on Intelligent Robots and Systems (IROS 2019), invited session chair of IEEE International Conference on Advanced Intelligent Mechatronics (AIM 2019), and PC member of IEEE International Conference on Robotics and Biomimetics (Robio 2019). He is currently serving as reviewer of many top journals and conferences, such as the *IEEE Transactions on Robotics*, the *IEEE/ASME Transactions on Mechatronics*, the *IEEE Transactions on Industrial Electronics*, the *IEEE Transactions on Neural Networks and Learning Systems*, the *IEEE Transactions on Circuits and Systems-I: Regular Papers, International Journal of Robust and Nonlinear Control*, ICRA, IROS, CDC, ACC, AIM. He is the senior member of IEEE.

Dr. Shengdong Yu, Wenzhou Institute, University of Chinese Academy of Sciences.

Prof. Shengdong Yu obtained his M.S. and Ph.D. degree from Mechanical and Electrical College of Nanjing University of Aeronautics and Astronautics. He is currently an associate professor of Electromechanical Engineering with the Wenzhou Institute, University of Chinese Academy of Sciences (WIUCAC). His current research interests include micro/nanosystems, micro/nanomechatronics, smart materials and structures, and sliding mode control theory. He has published a series of academic papers on the research of micro/nanomanipulation, such as, *ISA Transactions, Journal of the Franklin Institute, Bio Design and Manufacturing*, and *IEEE Transactions on Circuits and Systems I*.

1 Introduction

1.1 BACKGROUND

Cell micromanipulation (also known as cell microsurgery) refers to a microscopic biosurgical technique that completes the genetic manipulation of cells or embryos within the field of view of a biological microscope [1, 2]. Becoming one of the widely used technical means in biomedical engineering [3], cell micromanipulation mainly involves the following operations: positioning, nuclear transfer, cutting, and injection [4]. Figure 1.1 shows the three most frequently used cell micromanipulations.

Cell micromanipulation has been favored by medical scientists due to its irreplaceable superiority in physiology, pathology, and pharmacology and is mainly used in the following two fields of life science research:

1) As an important and widely used part of artificial reproduction technology, intracytoplasmic sperm injection uses cell micromanipulation to assist human reproduction [5, 6]. However, this far-reaching precision bioengineering still relies on tedious manual operations.
2) The genetic similarity between zebrafish and humans is as high as 70%. At the same time, with its advantages in genetics and highly conserved disease signaling trajectories [7], therefore, zebrafish is regarded as the preferred authentication bio for parsing. In recent years, scientists have carried out pharmacological, pathological, and reverse genetics studies through zebrafish and achieved specific gene mutation, overexpression, or deletion experiments in the zebrafish genome [8, 9]. By performing cell micromanipulation on a large number of zebrafish eggs, pharmaceutical and gene companies can monitor the development of zebrafish and complete large-scale studies of pharmacology and pathology.

Cells are tiny in size, usually only 10–100 microns in shape. For extremely fine microscopic operations such as cell injection or nuclear transfer, clinical personnel

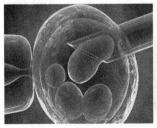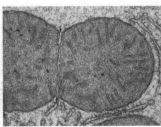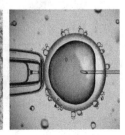

FIGURE 1.1 Three common micromanipulation techniques.

DOI: 10.1201/9781003294030-1

are required to have rich operating experience and can accurately penetrate the cell membrane of about 5 microns. Precision operation is achieved internally. However, mastering the above-mentioned micromanipulation skills not only requires long-term training of clinical personnel, but also frequent and boring cell operations will make clinical personnel prone to physical and psychological fatigue, which affects the quality and efficiency of cell manipulation. Therefore, manual micromanipulation, which is mainly artificial, severely restricts the development and progress of cell micromanipulation technology [10].

With the rapid development of precision manufacturing and automated control technology, automated cell micromanipulation in the microscopic field has been widely explored and practiced [11]. Figure 1.2 shows the traditional cell micromanipulation structure. The naked eye is observed through an inverted microscope, and the video signal of the electron microscope is connected to the computer for communication. In order to realize continuous automation, the cell arrays are distributed in containers, and the containers are placed on a mobile platform that can automatically feed at equal intervals [1]. During operation, the cells are guided to move to the lower part of the glass microneedle by machine vision technology, and the glass microneedle completes the automated operation of the cells. This automatic feeding method maintains the mechanical repetition of the cell injection action. Unfortunately, this method cannot realize the fine adjustment and subtle manipulation of the cell posture, and the cell survival rate is significantly reduced. At present, large-scale cell micromanipulation is still only at the conceptual research stage in the laboratory.

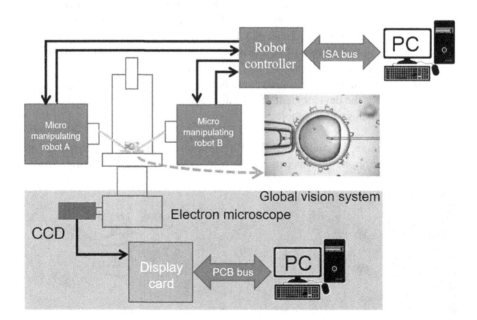

FIGURE 1.2   Functional framework of traditional cell micromanipulation system.

Therefore, the efficient acquisition of experimental data is severely restricted, and the reliability of experimental results is also greatly affected.

Therefore, it is urgent to improve the intelligent level of cell micromanipulation in a more intensive way and ensure good medical quality.

1.2 MOTIVATION

Cell micromanipulation is a kind of microbiological operation technology that can complete the gene operation of cells or embryos by using a biological microscope. This process involves various cell operations, such as localization, nuclear transfer, cutting, and injection. As the most widely used technology in cell micromanipulation, cell micropuncture has an irreplaceable medical value in the field of physiology, pathology, and pharmacology. In addition, biological cells with small size are usually very fragile, the operating instrument must have high precision to realize cell micropuncture with controlled puncture force.

This book gives a systematic description of advanced design, optimization, modeling, and hybrid force-position control techniques for positioning of cell puncture mechanism with piezoelectric actuator. Of particular interest, this book is devoted to the development of modeling and control methodologies for precise positioning of piezoactuated mechanism for biological cell puncture in the field of cell micromanipulation.

Researches on micro/nanorobots are mainly motivated by the advanced design, optimization, modeling, and control techniques of mechanical manipulation of biological cells from theoretical and experimental aspects. To achieve biomedical applications, reliability design, modeling, and precision control are vital to the development of engineering systems. With the advances in modeling, sensing, and control techniques, it is opportunistic to exploit them for the benefit of reliability design, actuation, and precision control of micro/nanomanipulation systems in order to expand the applications of robot at micro- and nanoscales, especially in biomedical engineering. The main focus of this book will be on the new techniques in design, modeling, actuation and robust control of piezoactuated mechanism for cell puncture.

1.3 SIGNIFICANCE

Cell micromanipulation technology occupies an irreplaceable key position in the development of biomedicine. Improving the automation degree of cell manipulation and survival rate necessitates the design, optimization, system identification, and hysteresis compensation of actuators, nonlinear robust control. Thus, comprehensive and in-depth research must be carried out on device design and other aspects.

With the development of new intelligent materials and control theory, this book presents high-resolution PEAs as the driving element and uses modern intelligent control methods as the support to develop a cell puncture mechanism with high motion accuracy. The position and force controls are comprehensively considered using a force-position hybrid controller based on multi-sensor information fusion that is constructed to achieve smooth switching between their subspaces. Tracking is carried out to control the deformation, thereby reducing damage to the cell.

1.4   CHAPTER ARRANGEMENT

The main chapters in this book are arranged as follows:

Chapter 1 systematically sorts and summarizes the global research status of micro/nano robots, cell manipulation technology, and micro-force sensors.

Chapter 2 discusses the optimized design of the parallel bridge displacement amplification mechanism. The first choice is to determine the type of flexible hinge suitable for cell puncture mechanism. Multi-objective optimization and sixth-order modal analysis are carried out using Ansys software to verify the correctness of the numerical calculations.

Chapter 3 introduces dynamic model and system identification of the cell puncture mechanism driven by a piezoelectric actuator, and the hysteretic nonlinearity of piezoelectric actuator is also analyzed.

Chapter 4 formulates a novel precise positioning control of a cell puncture mechanism driven by a piezoelectric actuator, considering compensation of the hysteretic nonlinearity of piezoelectric actuator. The strategy of feedforward control and sliding mode feedback control based on the Bouc–Wen inverse model is further developed to position the cell puncture mechanism. Proportional-integral sliding mode feedback control plus feedforward control has a simple structure and exhibits excellent performance.

Chapter 5 presents the optimized traditional Bouc–Wen model from the point of view of easy engineering application, and a nonlinear robust controller is constructed to integrate fractional non-singular terminal sliding mode and time-delay estimation (TDE) techniques. Then, the finite-time convergence and system stability are demonstrated using Lyapunov stability theory. The proposed controller is then verified through motion tracking experiments.

Chapter 6 introduces a novel precision robust motion controller for the completion of cell puncture experiments; the control strategy does not require prior knowledge of an unknown disturbance boundary. A sliding mode control (SMC) strategy with fast reaching law and proportional-integral-differential (PID)-type sliding surface based on the simplified Bouc–Wen model is combined with TDE technology to form an FPID-TDE controller. FPIDSMC and TDE combine and complement each other, with FPIDSMC reducing the burden of TDE and TDE reducing the gain of FPIDSMC.

Chapter 7 discusses the micro-force tracking control of the cell puncture mechanism, in preparation for the force-position mixing control in the next chapter. Based on the hysteresis dynamics model of the cell puncture mechanism, a micro-force tracking controller is designed. The TDE technology is used to estimate the unknown items, and the PID servo controller is used to achieve rapid convergence to errors and fast response to transient forces. Experiments show that the proposed controller can be effectively applied to micro-force tracking.

Chapter 8 synthesizes the theoretical and practical results of position and micro-force tracking controls in Chapters 5 and 6, respectively. An adaptive smooth switching algorithm for coupled force and position controls is proposed from the perspective of force-position hybrid control, which incorporates the FONTSM-based motion controller and the micro-force tracking controller for cell micropuncture into a unified control system. Then, using zebrafish embryo cells as the verification object, the force-position mixing control and compliant switching experiments are completed.

Chapter 9 presents a case study of cell micropuncture technique to achieve automated organelle biopsy of single cells with dimensions of less than 20 μm in diameter. The cells are patterned with a microfluidic device, and a template-matching-based image processing algorithm is developed to automatically measure the position of the desired organelles inside the cell. Followed by cell puncture manipulation, organelle extraction is then performed in an automatic way.

REFERENCES

1. Li YM, Tang H, Xu QS, et al. Development trend of micromanipulator robot technology for biomedical applications. *Journal of Mechanical Engineering* 2012; 47(23): 1–13.
2. Ghanbari A, Horan B, Nahavandi S, et al. Haptic microrobotic cell injection system. *Systems Journal IEEE* 2014; 8(2): 371–383.
3. Ali A, Abouleila Y, Shimizu Y, et al. Single-cell metabolomics by mass spectrometry: advances, challenges, and future applications. *TrAC Trends in Analytical Chemistry* 2019; 15(2): 1381–1400.
4. Xie M, Shakoor A, Shen Y, et al. Out-of-plane rotation control of biological cells with a robot-tweezers manipulation system for orientation-based cell surgery. *IEEE Transactions on Biomedical Engineering* 2019; 66(1): 199–207.
5. Yuan PQ, Zhang ZH, Luo S, et al. Application of computer-assisted semen analysis on sperm motility parameters in vitro fertilization. *Journal of Southern Medical University* 2013; 5(2): 448–450.
6. Luo YN. Clinical application of intracytoplasmic microinjection in oocytes of single sperm and analysis of related influencing factors. Fourth Military Medical University 2016.
7. Zuo P. Role and mechanism of glial maturation factor γ in zebrafish vascular bud cell migration and ovarian cancer invasion and metastasis. Zhejiang University 2016.
8. Peng CB. Study on traditional Chinese medicine compound and its drug target based on zebrafish drug screening platform. Guangzhou University of Traditional Chinese Medicine 2018.
9. Wang LJ. *Regulation of GABRB2 gene expression by DNA methylation in zebrafish: normal developmental regulation and dysregulation in a MeT-induced schizophrenia model*. Guangzhou, China: Southern Medical University, Guangzhou, 2016.
10. Xu D. *Microscopic Vision Measurement and Control*, Beijing: National Defense Industry Publication 2014.
11. Tan M, Wang S. Research progress in robotics. *Acta Automatica Sinica* 2013; 39 (7): 963–972.

2 Structural Design and Optimization of Cell Puncture Mechanism

2.1 INTRODUCTION

Figure 2.1 shows the overall structure of the constructed cell puncture system. The video signal of the electron microscope is connected to the computer for communication, and the operation is displayed on the screen. A fixation mechanism is used to maintain the cell position, and the puncture mechanism is used to penetrate the membrane. Compared with traditional motor drives, PEAs present the advantages of high displacement resolution and no transmission gap, electromagnetism, or noise pollution. Given their suitability for the precise operation of cells in a high-clean environment, PEAs are chosen as the driving element.

However, PEAs have a small range of motion [1], which severely limits its application range. This motion range can be doubled by using the displacement amplification mechanism. Among the micro–nano operating mechanisms, the compliant type [2–4] has the advantages of no transmission gap, no frictional resistance, high

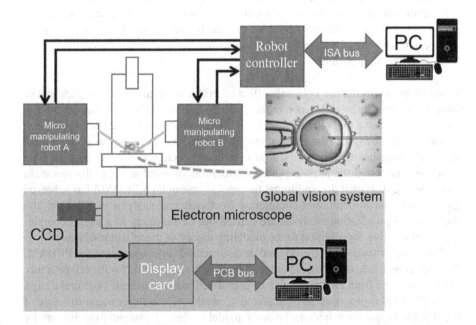

FIGURE 2.1 Overall structural design of the cell puncture system.

DOI: 10.1201/9781003294030-2

compatibility, and easy fabrication. The displacement amplifying mechanism is designed using the principle of the compliant mechanism, and then the PEAs are used to drive the parallel bridge displacement amplification mechanism (PBDAM) to obtain a large-stroke motion. Given that PEAs also have a backlash-free transmission, their mechanical properties highly match when embedded in the displacement amplifying mechanism, expanding the range of motion on the premise of maintaining accuracy.

Displacement scaling such as by lever magnification and Scott-Russell mechanisms [5] are widely used in the design of displacement magnification. In comparison, the PBDAM design based on the lever amplification mechanism has the characteristics of symmetrical mechanical properties and high magnification. A glass microneedle is attached to the output end of the PBDAM to ensure sufficient travel for cell manipulation. A flexible hinge is the core element of PBDAM to amplify displacement, which realizes the conversion of force and displacement through elastic deformation. However, the introduction of flexible hinges causes shortcomings in the PBDAM that need urgent solutions; the lateral stiffness and natural frequency are relatively low, and the flexible hinges are prone to mechanical damage.

Specifically, the issues are:

1) PEAs can only withstand compressive effects. Thus, if the PBDAM is not rigid enough to resist lateral loads, these may cause a potential threat to the failure of PEAs. To this end, a novel PBDAM is proposed in this study to improve the lateral stiffness.
2) A delicate and slender structure is often used to make the flexible hinge more prone to deformation, which then becomes the weakest link in the entire mechanism. Thus, we must not only obtain flexible motion effects but also to ensure sufficient strength and natural frequency, which requires the establishment of accurate mathematical models and optimization through intelligent algorithms.

Considering the hysteretic nonlinear effect of PEAs, we construct a complete static/ dynamic model of the cell puncture mechanism. For the design to obtain the maximum displacement magnification, checking the longitudinal and lateral stiffness, examining the stress distribution, and completing the strength check work are important. Literature has mainly focused on compliant mechanisms using rigid-flexible coupling theory and the pseudo-rigid body and the assumed mode as the two main methods [6]. Although the traditional theoretical research on PBDAM has achieved good results, an in-depth and comprehensive theoretical analysis of PBDAM and mechanism optimization methods, are carried out in this study as a useful complement and positive contribution to the modeling theory of compliant mechanisms.

This chapter focuses on the optimal design method for constructing the PBDAM. The first choice is to determine the type of flexible hinge suitable for the cell puncture mechanism, and from the perspective of lateral stiffness, the overall structural design of the PBDAM is proposed and completed. Considering the dimensional structure of the flexible hinge, we established a static model of the cell puncture mechanism by combining the pseudo-rigid body method [7] and the Euler-Bernoulli flexible beam theory [8]. The kinetic equation of the mechanism is obtained by the Lagrangian method [9] and then its natural frequency is obtained. The multi-objective optimization and sixth-order modal analysis of the cell puncture mechanism are carried out through the ANSYS Workbench simulation platform, and thus the numerical calculation and computer simulation are cross-validated.

2.2 OVERALL STRUCTURAL DESIGN OF THE CELL PUNCTURE MECHANISM

The PEAs are set at the center of the mechanism and its thrust is received at the input displacement of the PBDAM, which deforms and realizes the output displacement. The injection needle is fixed at the output displacement. A parallelogram arrangement of flexible hinge connections is used to ensure consistent needle orientation at all times. The displacement direction of PEAs is perpendicular to the injection direction of the micro-feed mechanism. The input of PBDAM is along the X-axis direction, and the movement direction of the injection needle is along the Y-axis direction, as shown in Figure 2.2. The base of the PEAs is fixed on the displacement amplifier through a locking mechanism. A spherical ceramic with a smooth surface is arranged on top of the PEAs, and the preloading mechanism applies a force of approximately 20 N to the spherical ceramic to fix the PEAs on the PBDAM and completely eliminate any gaps.

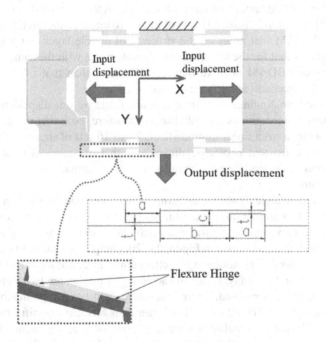

FIGURE 2.2 Partial enlarged view of PBDAM and flexible hinge. (Source: Ma, S. Yu, S. Kang, Y. Shen et al./Journal of Agricultural Machinery, 2021, 52 (09): 417–426, with permission.)

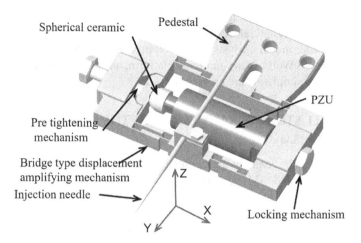

FIGURE 2.3 Virtual prototype design of the cell puncture mechanism. (Source: Ma, S. Yu, S. Kang, Y. Shen et al./Journal of Agricultural Machinery, 2021, 52 (09): 417–426, with permission.)

A significant disadvantage of conventional PBDAM is the relatively low lateral stiffness [10], which affects the lifetime of PEAs. To improve the lateral stiffness, we construct a PBDAM that relies on the structure of double-layer composite hinges. Figure 2.3 shows that for the straight beam flexible hinge with the same size the lateral stiffness of PBDAM is more than 38.6% higher than that of the traditional bridge-type displacement amplifying mechanism.

At the same time, findings show that: 1) A set of side-by-side displacement amplifying mechanisms constitutes a parallelogram structure, such that the displacement output can obtain a strict linear motion effect; 2) All 16 sets of straight beam flexible hinges can be consistent geometric dimensions that are adopted to simplify the design process while ensuring the symmetry of the mechanical properties of the overall structure.

Circular and straight beam type flexure hinges are widely used in compliant mechanisms [11]. Circular arc flexure hinges [12] are generally considered to have high swing accuracy, and the central axis hardly drifts during swinging while at the original position float, as shown in Figure 2.4. By comparison, straight beam flexure hinges have a larger deformation range, better compliance, and more DOF of motion (see Figure 2.5). In this study, range and accuracy of motion are two technical indicators that cannot be separated. Therefore, the straight beam type flexible hinge is used in the design of PBDAM to obtain a larger displacement magnification. Motion accuracy is collected by a high-precision laser displacement sensor, and the closed-loop feedback control is performed by a nonlinear robust controller. Thus, the cell puncture mechanism can obtain a large stroke and high precision manipulation capability.

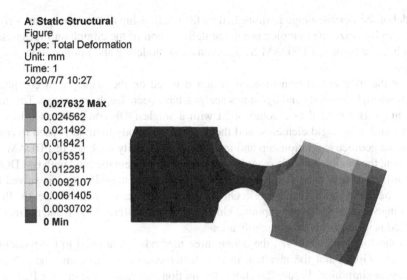

FIGURE 2.4   Simulation diagram of bending deformation of arc flexible hinge. (Source: Ma, S. Yu, S. Kang, Y. Shen et al./Journal of Agricultural Machinery, 2021, 52 (09): 417–426, with permission.)

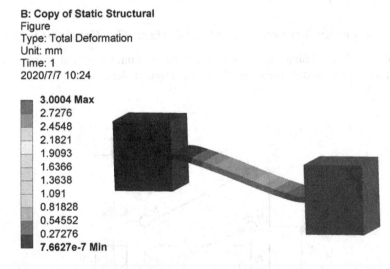

FIGURE 2.5   Simulation diagram of bending deformation of straight beam flexible hinge. (Source: Ma, S. Yu, S. Kang, Y. Shen et al./Journal of Agricultural Machinery, 2021, 52 (09): 417–426, with permission.)

2.3   OVERALL STATIC ANALYSIS

Professors Midha and Howell proposed a pseudo-rigid body model [10], established the "force-deformation" criterion in compliant mechanisms, and devised an important research method for rigid-flexible coupling. For this reason, the mechanical model of the flexible hinge is simplified by the relationship of "torsion spring + rigid link", to linearize the complex nonlinear deformation of the compliant mechanism, which is the theory of PBDAM for mathematical modeling and optimization design support.

For the mechanical transmission designed based on the principle of compliant mechanism, kinematics and dynamics analysis have been deeply examined. The flexible hinge is regarded as a rotary joint with a single DOF composed of a torsion spring and other rigid elements, and the pseudo-rigid body method is used to complete the geometric relationship and instantaneous velocity analyses of PBDAM. In addition, the flexible hinge is regarded as a compound kinematic joint with two DOFs of tensile and rotational stiffness, and the displacement magnification is solved by using the principle of virtual work. On this basis, considering its local size, the flexible hinge is regarded as a compound kinematic joint with three DOFs, and the matrix method is used to solve the magnification.

In the following analysis, the above three methods are applied to the modeling PBDAM. Given that the mechanism is a center-symmetric structure, the solution process is simplified. Figure 2.6 shows the motion diagram of the quarter PBDAM, where the quarter model is used as the research object. Input force and displacement are applied to the quarter mechanism, respectively, and comprise the output displacement. Therefore, the input force, input displacement, and output displacement are needed for a complete mechanism.

2.3.1 MODELING ASSUMPTIONS FOR FLEXIBLE HINGE ONE DEGREE OF FREEDOM

Consider the flexible hinge as a single DOF rotation joint, the axis of rotation is the central part of the flexible hinge and the rest are rigid bodies.

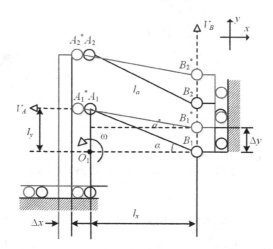

FIGURE 2.6 Motion diagram of a quarter PBDAM. (Source: Ma, S. Yu, S. Kang, Y. Shen et al./Journal of Agricultural Machinery, 2021, 52 (09): 417–426, with permission.)

Figure 2.6 shows that PEAs generate a horizontal input displacement, which causes the flexible hinge to deform, and causes the output displacement of PBDAM. At the same time, the inclination of the upper or the lower link is converted from α to α^*.

$$l_a \cos\alpha + \Delta x = l_a \cos\alpha^* \tag{2.1}$$

$$l_a \sin\alpha - \Delta y = l_a \sin\alpha^* \tag{2.2}$$

The two equations above denote the lengths of the upper and lower links. Take Squares (2.1) and (2.2) and add them together to eliminate α^*:

$$\Delta y^2 - 2l_a \sin\Delta y + \Delta x^2 + 2l_a \cos\Delta x = 0 \tag{2.3}$$

After finishing Equation (2.3), the output displacement Δy can be obtained:

$$\Delta y = l_a \sin\alpha - \sqrt{l_a^2 \sin^2\alpha - \Delta x^2 - 2l_a \Delta x \cos\alpha} \tag{2.4}$$

Therefore, the displacement magnification of the mechanism can be calculated as:

$$C_1 = \frac{\Delta y}{\Delta x} = \frac{l_a \sin\alpha}{\Delta x} - \sqrt{\frac{l_a^2 \sin^2\alpha - \Delta x^2 - 2l_a\Delta x\cos\alpha}{\Delta x}} \tag{2.5}$$

The above formula shows that for the specified mechanism, and are given, therefore, the displacement magnification only shows a nonlinear relationship with the input displacement.

Equation (2.2) shows the maximum output displacement of the mechanism, that is, when the limit value of Δy occurs at $\alpha^* \to 0$, then $\Delta y_{\max} \to l_a \sin\alpha$.

2.3.2 2-DOF MODELING ASSUMPTIONS FOR FLEXIBLE HINGES

The flexible hinge has tensile and rotational stiffness, and the rest is a rigid body. Figure 2.7 shows the force state for one of the links, such as the upper link. The moment equation is established at point A, and the following formula is obtained.

$$F_x l_a \sin\alpha = 2M_r = 2K_r\Delta\alpha \tag{2.6}$$

Among them, the moment is produced by the action of the rotational stiffness, causing the angle of rotation of the upper link.

Establish a force equation for point A as

$$F_l = \frac{F_x}{\cos\alpha} = K_t\Delta l \tag{2.7}$$

where is the axial displacement due to tensile stiffness.

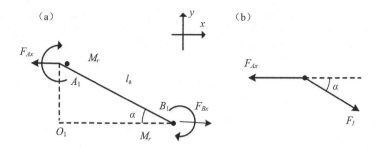

FIGURE 2.7 Force diagram of a connecting rod in a quarter PBDAM. (Source: Ma, S. Yu, S. Kang, Y. Shen et al./Journal of Agricultural Machinery, 2021, 52 (09): 417–426, with permission.)

Based on the principle of virtual work, the following energy equation is constructed,

$$F_x \Delta x = F_l \Delta l + 2 M_r \Delta \alpha \qquad (2.8)$$

Combining the above three equations, we obtain

$$F_x \Delta x = K_t \Delta l^2 + 2 K_t \Delta \alpha^2 \qquad (2.9)$$

Then, we construct the relationship between the input displacement and the input force,

$$\Delta x = \frac{2 K_r + K_t l_\alpha^2 \cos^2 \alpha \sin^2 \alpha}{2 K_t K_r \cos^2 \alpha} F_x \qquad (2.10)$$

Taking the derivative of the equation $l_y = l_a \sin \alpha$ with respect to time, we obtain

$$\Delta y = l_\alpha \cos \alpha \, \Delta \alpha \qquad (2.11)$$

The displacement magnification can be obtained as

$$C_2 = \frac{\Delta y}{\Delta x} = \frac{K_t l_\alpha^2 \cos^3 \alpha \sin \alpha}{2 K_r + K_t l_\alpha^2 \cos^2 \alpha \sin^2 \alpha} \qquad (2.12)$$

For the overall mechanism, its input stiffness can be obtained as

$$K_{in} = \frac{2 F_x}{\Delta x} = \frac{4 K_t K_r \cos^2 \alpha}{2 K_r + K_t l_\alpha^2 \cos^2 \alpha \sin^2 \alpha} \qquad (2.13)$$

2.3.3 RIGID-FLEXIBLE COUPLING MODELING OF FLEXIBLE HINGES WITH VARIABLE SECTIONS

The modeling and analysis of the flexible hinge above has been completed, and the displacement magnification is also obtained. However, the local size of the flexible hinge is not involved, which is clearly insufficient for precise analysis. Next, the detailed size of the flexible hinge is considered, and the accurate static model of the PBDAM is established by combining the pseudo-rigid body method and Euler-Bernoulli flexible beam theory. Subsequently, the dynamic equation of the mechanism is established according to the Lagrange method to finally obtain the natural frequency of the mechanism.

Figure 2.8 shows that the output displacement Δy of the mechanism mainly comes from the bending deformation of the upper link AB and the lower link CD. To establish the moment equation at point A, we obtain

$$F_x l_y - 2M_r = 0 \tag{2.14}$$

Given that the flexible hinge is variable, the upper link AB or the lower link CD is divided into three sections, the heights of which are respectively

$$h_{1,2}(d_x) = t, d_x \in [0,a] \cup [a+b, 2a+b] \tag{2.15}$$

$$h(d_x) = c, d_x \in (a, a+b) \tag{2.16}$$

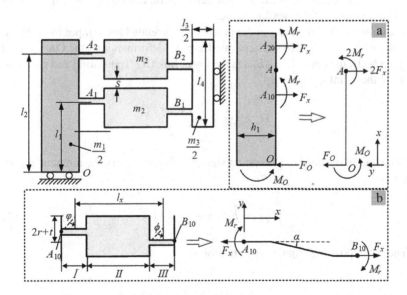

FIGURE 2.8 Force diagram of a quarter PBDAM considering the local dimensions of the hinge. (Source: Ma, S. Yu, S. Kang, Y. Shen et al./Journal of Agricultural Machinery, 2021, 52 (09): 417–426, with permission.)

The torques of the three stages are

$$M_1(d_x) = M_r, d_x \in [0, a] \tag{2.17}$$

$$M_2(d_x) = M_r - \frac{F_x(d_x - a)l_y}{l_x}, d_x \in (a, a + b) \tag{2.18}$$

$$M_3(d_x) = -M_r, d_x \in [a + b, 2a + b] \tag{2.19}$$

According to Euler-Bernoulli flexible beam theory, the angular deformation of the three segments of the flexible hinge is established as

$$\theta_1(\varphi) = \int \frac{12M_1(\varphi)}{Ewh_1^3(\varphi)} r\cos\varphi d\varphi + P_1 \tag{2.20}$$

$$\theta_2(x) = \int \frac{12M_2(\varphi)}{Ewh_2^3(x)} dx + P_2 \tag{2.21}$$

$$\theta_3(\phi) = \int \frac{12M_3(\varphi)}{Ewh_3^3(\phi)} r\cos\phi d\phi + P_3 \tag{2.22}$$

where $P_i (i = 1, 2, 3)$ is constant.

The integral operation of the angular deformation can obtain the displacement of the three segments of the flexible hinge.

The input displacement Δx of the mechanism is related to the input force F_x, given that the quarter model shows Δx coming from the deformation of the OA segment of the main body, the upper link AB, and the lower link CD. Translating and merging the forces at these links, we obtain

$$M_0 + 2M_r = 2F_x \frac{l_1 + l_2}{2} \tag{2.23}$$

Referring to the coordinate system shown in Figure 2.8(a), the moment at coordinate x is

$$M(x) = F_x(2x - l_1 - l_2 + l_y) \tag{2.24}$$

The corresponding angular deformation is

$$\theta(x) = \frac{12F_x(x^2 - l_1x - l_2x + l_yx)}{Ewh_1^3} + P_4 \tag{2.25}$$

Here, P_4 is a constant.

Computing the integral operation on the above formula, the displacement deformation amount is

$$y(x) = \frac{2F_x\left(2x^3 - 3l_1x^2 - 3l_2x^2 + 3l_yx^2\right)}{Ewh_1^3} \tag{2.26}$$

Therefore, the displacement deformation at point A is

$$\Delta x_1 = y\left(\frac{l_1 + l_2}{2}\right) \tag{2.27}$$

Figure 2.9 shows that the deformation of the upper link *AB* or the lower link *CD* due to tensile deformation is $\Delta L = 2F_x/K_r$. The displacements along the y- and x-axes are Δx_2 and Δy, respectively, and are calculated as follows:

$$\Delta x_2 = (L_\alpha + \Delta L)\cos\alpha^* - L_\alpha \cos\alpha \tag{2.28}$$

$$\Delta y = L_\alpha \sin\alpha - (L_\alpha + \Delta L)\sin\alpha^* \tag{2.29}$$

where $L_\alpha = \sqrt{(l_x + 2r)^2 + l_y^2}$ is the approximate length of the connecting rod.

After eliminating α^*, Δx_2 can be obtained as

$$\Delta x_2 = \sqrt{(L_\alpha + \Delta L)^2 - (\Delta y - L_\alpha \sin)^2} - L_\alpha \cos\alpha \tag{2.30}$$

Thus far, the displacement magnification of the mechanism can be obtained as

$$C_3 = \frac{\Delta y}{\Delta x} = \frac{\Delta y}{\Delta x_1 + \Delta x_2} \tag{2.31}$$

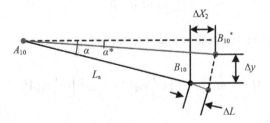

FIGURE 2.9 Schematic diagram of the deformation process of the upper link AB due to tensile deformation. (Source: Ma, S. Yu, S. Kang, Y. Shen et al./Journal of Agricultural Machinery, 2021, 52 (09): 417–426, with permission.)

The input stiffness of the mechanism is

$$K_{in} = \frac{2F_x}{\Delta x} \tag{2.32}$$

In summary, the displacement magnification and input stiffness of the mechanism are established through three model assumptions.

2.4 NATURAL FREQUENCY ANALYSIS BASED ON LAGRANGIAN METHOD

For PBDAM, the input displacement is $q = 2\Delta x$ and the kinetic and potential energies of the mechanism are

$$
\begin{aligned}
T &= \frac{1}{2}m_3 d^2 + 2 \times \frac{1}{2}m_1 \left[\left(\frac{\dot{q}}{2}\right)^2 + \left(\frac{\dot{d}}{2}\right)^2 \right] + 8 \times \frac{1}{2} \times \frac{1}{12} m_2 l_\alpha^2 \dot{\theta}^2 \\
&+ 4 \times \frac{1}{2}m_2 \left[\left(\frac{\dot{q}}{4}\right)^2 + \left(\frac{\dot{d}}{4}\right)^2 \right] + 4 \times \frac{1}{2}m_2 \left[\left(\frac{\dot{q}}{4}\right)^2 + \left(\frac{3\dot{d}}{4}\right)^2 \right]
\end{aligned} \tag{2.33}
$$

$$V = 2K_r \left(\frac{d}{l_a}\right)^2 \tag{2.34}$$

Substitute the above energy formula into the following Lagrangian equation,

$$\frac{d}{dt} \cdot \frac{\partial T}{\partial \dot{q}} - \frac{\partial T}{\partial q} + \frac{\partial V}{\partial q} = F_{in} \tag{2.35}$$

The above formula can be organized as follows,

$$M\ddot{q} + Kq = 0 \tag{2.36}$$

where the equivalent mass and stiffness coefficients are

$$M = \frac{1}{4}m_1\left(1+A^2\right) + \frac{1}{4}m_2\left(1+\frac{31}{3}A^2\right) + \frac{1}{2}m_3 A^2 \tag{2.37}$$

$$K = \frac{2A^2 K_r}{l_a^2} \tag{2.38}$$

Therefore, the natural frequencies of institutions are

$$f_n = \frac{\omega}{2\pi} = \frac{1}{2\pi}\sqrt{\frac{K}{M}} = \frac{A}{\pi l_a}\sqrt{\frac{2K_r}{m_1\left(1+A^2\right)+m_2\left(1+\frac{31}{3}A^2\right)+2m_3A^2}} \tag{2.39}$$

2.5 OPTIMIZATION OF THE GEOMETRIC DIMENSIONS OF THE MECHANISM

Based on the previously established mathematical model of PBDAM, this section completes the geometric optimization of the mechanism through numerical calculations. The intelligent optimization algorithm takes the maximum value of the natural frequency as the optimization goal to improve efficiency, and determines the geometric size of the mechanism under the constraints of relevant boundary conditions. This is a feasible and convenient method to carry out finite element simulation analysis through ANSYS Workbench to mutually verify the correctness of numerical optimization and computer simulation.

The structure of PBDAM is delicate and complex, and 3D printing technology provides a convenient way for its preparation. The material is a mixture of photosensitive resin and polypropylene, with a geometric accuracy of ±0.015 mm. The material properties are density: 905 Kg/m3, Young's modulus: 1.5×109 Pa, and Poisson's ratio: 0.41.

2.5.1 GEOMETRIC DIMENSION OPTIMIZATION BASED ON DIFFERENTIAL EVOLUTION ALGORITHMS

Given that the rigid-flexible coupling modeling of variable cross-section more precisely reflects the geometric size of the flexible hinge, which includes the core content of optimization, the displacement amplification is established based on the rigid-flexible coupling modeling of variable cross-section. The multiplier, input stiffness, and the natural frequency of the mechanism is also used in the optimization.

The differential evolution (DE) algorithm is formed based on the random parallel search strategy of groups, to simulate the survival mode of cooperation and competition among various groups in nature. The calculation is more simplified and efficient than the traditional genetic algorithm. Due to the complexity of this optimization calculation, a factor that changes with the increase of the population is used to improve the convergence speed of the algorithm. The ideas for the DE algorithm include:

(1) Population initialization. The selection of the initial population must come from the entire solution space.
(2) Differential variation. The purpose of this step is to enhance the global search ability of the algorithm, and the optimal value of the solution space can be obtained through the DE/rand/1 strategy.

(3) Crossover operation. In the subsequent step of differential mutation, by setting the crossover probability CR as a dynamic variable that changes with
the fitness function, the local search ability can also be considered on the
premise of ensuring global search ability, the. This step is the core of the
algorithm.
(4) Select an operation. Individuals with large fitness are eliminated according
to the greedy strategy.

The above is the main content of the DE algorithm. Figure 2.10 shows the four steps
of the DE algorithm.

To obtain a compact mechanical structure, we use the four parameters (a, b, c, t)
of the flexible hinge as the main design variables, for which a more reasonable value
range is set. Therefore, this range evolves to a single-objective optimization problem
under multiple constraints. Optimization calculations need to be carried out under
the boundary conditions stated in Tables 2.1 and 2.2.

After DE optimization, the dimensions of the flexible hinge are: a (3.1252mm), b
(6.4608mm), c (4.3552mm), and t (0.2905mm), and the natural frequency is
201.68Hz.

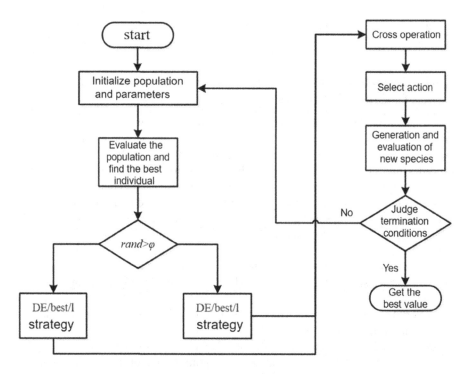

FIGURE 2.10   DE algorithm flow chart. (Source: Ma, S. Yu, S. Kang, Y. Shen et al./Journal
of Agricultural Machinery, 2021, 52 (09): 417–426, with permission.)

TABLE 2.1

Constraints and Optimization Objectives

	Restrictions	optimize the target
output displacement	8 times greater than the input displacement	
Safety factor	greater than 10	
Enter stiffness	Less stiffness than PEAs	
natural frequency	Greater than 200Hz	maximum value

Source: Ma, S. Yu, S. Kang, Y. Shen et al./Journal of Agricultural Machinery, 2021, 52 (09): 417–426, with permission.

TABLE 2.2

Value Range of Flexible Hinge Design Variables

variable symbol	design variable	Ranges (mm)
a	Length of flexible beam	2–5
b	length of rigid beam	5–10
c	Thickness of rigid beam	1–5
t	Thickness of flexible beam	0.25–0.5

Source: Ma, S. Yu, S. Kang, Y. Shen et al./Journal of Agricultural Machinery, 2021, 52 (09): 417–426, with permission.

2.5.2 GEOMETRY OPTIMIZATION BASED ON FINITE ELEMENT SIMULATION TECHNOLOGY

After the necessary simplification of the model, the geometry construction is completed in ANSYS Workbench 19.0 software, and the design variables are parameterized. After the parts are divided into several slices, they are merged into one and yield 95569 nodes and 19524 hexahedral elements. Loads and set for the model are configured with two input displacements of five microns each. Static and then modal analyses are carried out to correlate the natural frequency of the part with the main design variables. The optimization calculation is completed with the help of the multi-island genetic algorithm in the ANSYS Workbench 19.0 software toolbox. Table 2.3 shows the results. Considering the limitation of manufacturing precision, the actual dimensions of the flexible hinge are a (3.13mm), b (6.46mm), c (4.35mm), t (0.29mm).

2.5.3 SIXTH-ORDER MODAL ANALYSIS AND VERIFICATION OF OPTIMIZATION RESULTS

The results of static analysis reflect the strain state and stress distribution of the structure. Figure 2.11 shows that the maximum output displacement of the mechanism is

TABLE 2.3

Results of Optimization Using Finite Element Simulation

	result 1	result 2	result 3
a(mm)	3.1262	3.1782	3.2287
b(mm)	6.4658	6.4671	6.4654
c(mm)	4.3526	4.3522	4.5002
t(mm)	0.28995	0.29094	0.28492
displacement output(μm)	89.916	89.883	89.996
Displacement magnification	8.99	8.98	8.99
Enter stiffness(N/m)	6.61×10^3	6.51×10^3	6.49×10^3
Safety factor	15	15	15
natural frequency(Hz)	213.15	212.88	212.66

Source: Ma, S. Yu, S. Kang, Y. Shen et al./Journal of Agricultural Machinery, 2021, 52 (09): 417–426,
with permission.

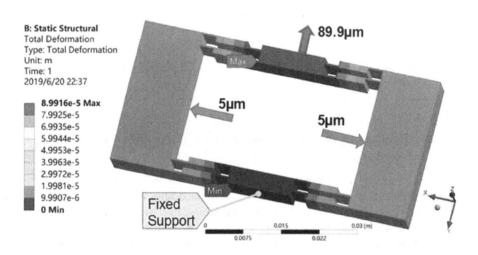

FIGURE 2.11 Deformation cloud diagram with five microns of deformation applied to both sides. (Source: Ma, S. Yu, S. Kang, Y. Shen et al./Journal of Agricultural Machinery, 2021, 52 (09): 417–426, with permission.)

89.9 μm when a displacement of 5 μm is applied to both sides; Figure 2.12 shows that when a displacement of 10 μm is applied to both sides of the mechanism, the maximum stress is 1 MPa. The modal simulation analysis of the structure shows that the first-order resonance frequency is 41.91 Hz, which is high enough for cell puncture.

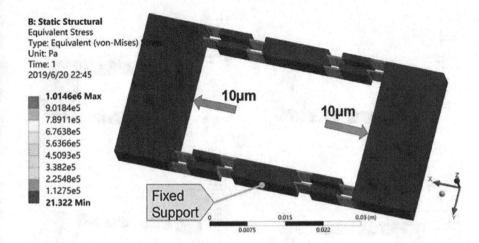

FIGURE 2.12 Stress cloud diagram after applying a load of 10 microns to both sides. (Source: Ma, S. Yu, S. Kang, Y. Shen et al./Journal of Agricultural Machinery, 2021, 52 (09): 417–426, with permission.)

TABLE 2.4
First Six Modes of PBDAM

Modes	Frequency (Hz)
1	41.91
2	135.82
3	160.85
4	328.75
5	387.11
6	708.05

Source: Ma, S. Yu, S. Kang, Y. Shen et al./
Journal of Agricultural Machinery,
2021, 52 (09): 417–426, with
permission.

Table 2.4 shows the sixth-order natural frequencies of PBDAM. The natural frequency is increased by 5.39% compared with DE optimization, which proves that the mathematical model and optimization of the flexible hinge have reliable accuracy. Thus, the two research approaches achieve the purpose of mutual verification.

Figure 2.13 shows the first six-order modal deformation diagram of PBDAM. The first-order natural frequency mainly comes from the outer link of the displacement output platform (shown in red), while the second-order natural frequency mainly comes from the right side of the mechanism (shown in red).

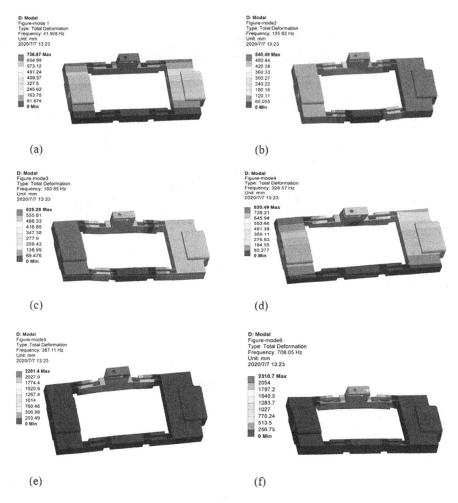

FIGURE 2.13 First six modal deformation diagram of PBDAM. (a) First order modal cloud diagram. (b) Second order modal cloud diagram. (c) Third order modal cloud diagram. (d) Fourth order modal cloud diagram. (e) Fifth order modal cloud diagram. (f) Sixth order modal cloud diagram. (Source: Ma, S. Yu, S. Kang, Y. Shen et al./Journal of Agricultural Machinery, 2021, 52 (09): 417–426, with permission.)

2.6 CONCLUSION

This chapter mainly completes the mechanical structure design of the PBDAM. The main results are as follows:

(1) PBDAM is proposed to improve the lateral stiffness of the bridge-type displacement amplifying mechanism, and make PEAs achieve a better stress state. At the same time, the upper and lower links are in a parallelogram structure to ensure that the direction of the injection needle moves strictly in a straight line along the a-axis.

(2) According to rigid-flexible coupling theory, the PBDAM is deeply studied and three model assumptions are established for the flexible hinge: a rotating joint with a single DOF, a composite kinematic joint with two DOFs with tensile and rotational stiffness while considering flexibility. A variable-section rigid-flexible coupled model of hinge local dimensions. The dynamic model of PBDAM is obtained by using Lagrange method to determine the natural frequency of the mechanism.

(3) The geometric size of the PBDAM is optimized with the maximum value of the natural frequency using DE algorithm. The multi-objective optimization and sixth-order modal analysis of the cell puncture mechanism are carried out by finite element software. Comparative analysis is further accomplished showing that the mathematical model of PBDAM has high accuracy, which illustrate the effectiveness of the proposed design principle and optimization method.

REFERENCES

1. Liu Y, Shi S, Yan J, et al. A novel piezoelectric actuator with two operating modes. *Journal of Intelligent Material Systems and Structures* 2018; 29(6): 1157–1164.
2. Lazarov BS, Schevenels M, Sigmund O. Robust design of large-displacement compliant mechanisms[J]. *Mechanical Sciences* 2011; 2(2): 175–182.
3. Pedersen CBW, Buhl T, Sigmund O. Topology synthesis of large-displacement compliant mechanisms. *International Journal for Numerical Methods in Engineering* 2015; 50(12): 2683–2705.
4. Howell LL, Midha A. A loop-closure theory for the analysis and synthesis of compliant mechanisms. *ASME Journal of Mechanical Design* 1996; 118(1): 121–125.
5. Tian Y, Shirinzadeh B, Zhang D, et al. Development and dynamic modelling of a flexure-based Scott-Russell mechanism for nano-manipulation. *Mechanical Systems & Signal Processing* 2009; 23(3): 957–978.
6. Jia-Hao C, Zhi-Qiang HU, Ge-Liang L, et al. Study on rigid-flexible coupling effects of floating offshore wind turbines. *China Marine Engineering* 2019; 33(1): 689–702.
7. Pei X, Yu J, Zong G, et al. An effective pseudo-rigid-body method for beam-based compliant mechanisms. *Precision Engineering* 2015; 34(3): 634–639.
8. Zhigang Z, Zhaohui Q, Zhigang W, et al. A spatial Euler-Bernoulli beam element for rigid-flexible coupling dynamic analysis of flexible structures. *Shock and Vibration* 2015; 5(12): 1–15.
9. Gong X, Takagi S, Huang H, et al. A numerical study of mass transfer of ozone dissolution in bubble plumes with an Euler-Lagrange method. *Chemical Engineering ENCE* 2007; 62(4): 1081–1093.
10. Howell LL, Midha A. A method for the design of compliant mechanisms with small-length flexural pivots. *Journal of Mechanical Design* 1994; 116(1): 280–290.
11. Xu W, King T. Flexure hinges for piezoactuator displacement amplifiers: flexibility, accuracy, and stress considerations. *Precision Engineering* 1996; 19(1): 4–10.
12. Yong YK, Lu TF, Handley DC. Review of circular flexure hinge design equations and derivation of empirical formulations. *Precision Engineering* 2008; 32(2): 63–70.

3 Dynamic Modeling, System Identification, and Hysteresis Effect of the Cell Puncture Mechanism

3.1 DYNAMIC MODELING

This chapter describes the study of the dynamic model of the entire cell puncture mechanism, including PEA and bridge-type displacement amplification mechanism, instead of focusing solely on a single PEA. The dynamic model of the entire micro-puncture mechanism is constructed according to the principle of the Bouc–Wen model [1, 2]. The mechanism consists of second-order linear and nonlinear components of PEA, and the entire dynamic model including hysteretic nonlinearity is established as follows:

$$M\ddot{x} + C\dot{x} + k_e x = k_e\left(du - h\right) + f_d \tag{3.1}$$

$$\dot{h} = \xi_1 d\dot{u} - \xi_2 \left|\dot{u}\right| h - \xi_3 \dot{u}\left|h\right| \tag{3.2}$$

Parameters M, C, k_e, and x represent the equivalent mass, damping coefficient, stiffness coefficient, and displacement, respectively. In addition, d is the piezoelectric coefficient, u is the input voltage, f_d is the total disturbance of the unmodeled terms and external disturbances, and h is the hysteresis effect of the system. ξ_1, ξ_2, and ξ_3 are the coefficients describing the shape of the hysteresis loop.

The Bouc–Wen model can describe only symmetric hysteresis loops. In practice, the hysteresis loop is often asymmetric, which is the disadvantage of the Bouc–Wen model. Although the modeling is incomplete, it can be offset by improving the performance of robust controller.

3.2 HYSTERESIS EFFECT

The relationship between the input voltage and output displacement is described through an open-loop experiment, and the hysteresis effect is observed. Figure 3.1 (a) shows that a sinusoidal signal with gradually increasing frequency is applied to the cell puncture mechanism to obtain the relation curve between the input voltage

DOI: 10.1201/9781003294030-3

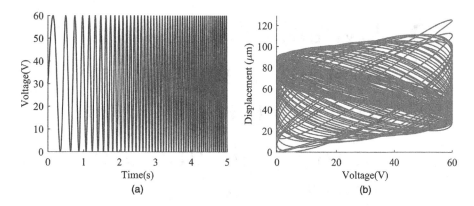

FIGURE 3.1 Hysteresis effect of the cell puncture mechanism: (a) Amplitude of input voltage is 120 V. The frequency gradually increases; (b) Curve of input voltage and output displacement of the cell puncture mechanism. (Source: M. Xie, S. Yu, H. Lin et al. /IEEE Transactions on Circuits and Systems I: Regular Papers 67 (9) 3199-3210, 2022, with permission.)

and output displacement. Figure 3.1 (b) shows that the input voltage and output displacement are not linear and the hysteresis loop changes continuously. At the same time, the magnitude and direction of the hysteresis loop change with the increase in input voltage frequency. In particular, the magnitude of the hysteresis effect is related to the input voltage frequency.

3.3 SYSTEM IDENTIFICATION

3.3.1 IDENTIFICATION OF LINEAR COMPONENT

The previous analysis [3] indicates that hysteretic-parameter identification can be avoided by classifying hysteresis and disturbance into unknown terms. A small-amplitude sinusoidal voltage (e.g., 5 V) can be applied to the micropuncture mechanism to complete the identification of the linear part. In this case, the hysteresis effect is highly suppressed and can be omitted. The left term of the dynamic model (Eq. (3.1)) is proportional to the input voltage. The model is approximately a linear system. Accordingly, the output force is a linear force proportional to the input voltage.

$$M\ddot{x} + C\dot{x} + k_e x = k_e du \qquad (3.3)$$

Through the Laplacian transformation of Eq. (3.3), the following input and output transfer functions can be obtained:

$$\frac{X(s)}{U(s)} = \frac{k_e d}{Ms^2 + Cs + k_e} \qquad (3.4)$$

M is obtained by electronic balance. A small sweep sinusoidal driving voltage is applied to PEA, and the frequency response diagram is drawn by the measured data.

The parameters C, k_e, and d can be identified accurately by using the system identification toolbox of MATLAB. Accordingly, we obtain the following parameters: $M = 2.3 \times 10^{-2}$ kg, $C = 1.6 \times 10^3$ Ns/m, $k_e = 1 \times 10^4$ N/m, and $d = 9.8 \times 10^{-7}$ m/V.

3.3.2 IDENTIFICATION OF HYSTERETIC NONLINEAR COMPONENT

Different frequencies of the input signal exert an obvious influence on the cell puncture mechanism. In the identification of the hysteresis component, the input low-frequency voltage signal can reduce the influence of frequency. Particle swarm optimization (PSO) algorithm is used to identify hysteresis loop coefficients, namely, ξ_1, ξ_2, and ξ_3 in Eq. (3.2). Without losing generality, we use the conventional objective function (i.e., root mean square error) as the cost function to describe the accuracy of identification.

$$J\left(\xi_1, \xi_2, \xi_3\right) = \sqrt{\frac{1}{N} \sum_{i=1}^{N} \left(x_h(i) - x(i)\right)^2} \tag{3.5}$$

where $x_h(i)$ is the output displacement of the cell puncture mechanism measured at time i, $x(i)$ is the output displacement of the model at time i, and N is the total number of samples.

PSO is a type of evolutionary algorithm similar to the genetic algorithm. PSO searches for the optimal solution by iteration and evaluates the quality of the solution by cost function. PSO has simpler rules, easier implementation, higher accuracy, and faster convergence than the genetic algorithm. Thus, PSO is widely used in function optimization and system identification.

The process of parameter identification is as follows. First, sinusoidal AC voltage is input to PEA, wherein the amplitude is 0–80 V, the frequency is 1 Hz, and the sampling time is set to 0.0001 s. Second, the displacement curve is obtained. Third, hysteresis loop coefficients are identified with PSO algorithm. The minimum value

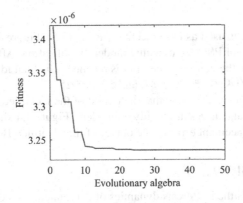

FIGURE 3.2 Convergence procedure of PSO for dynamic model identification. (Source: S. Yu, M. Xie, H. Wu et al. /ISA Transactions 124 (2022) 427–435, with permission.)

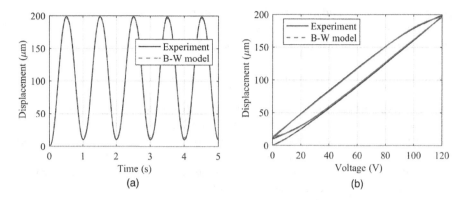

FIGURE 3.3 Displacement output of the identified Bouc–Wen model compared with the experimental data: (a) time–displacement curve and (b) voltage–displacement curve. (Source: S. Yu, M. Xie, H. Wu et al. /ISA Transactions 124 (2022) 427–435, with permission.)

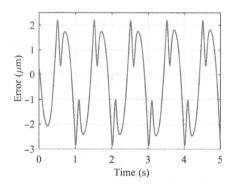

FIGURE 3.4 Displacement error curve of the identified Bouc–Wen model output compared with the experimental data. (Source: S. Yu, M. Xie, H. Wu et al. /ISA Transactions 124 (2022) 427–435, with permission.)

of the cost function is used as the objective via Eq. (3.5). Figure 3.2 shows the convergence procedure of PSO for dynamic model identification. After optimization in the 20th generation, the convergence limit is reached. The final identification results are as follows: $\xi_1 = 0.4588$, $\xi_2 = 0.0587$, and $\xi_3 = 0.0025$.

Figures 3.3(a) and 3.3(b) show that the output of the identified Bouc–Wen model and experimental data curves are highly coincident. Figure 3.4 shows that 2.8 μm is the maximum displacement error of the output of the identified Bouc–Wen model.

3.4 CONCLUSION

This chapter models the hysteresis dynamics of cell puncture mechanism. The main contributions of this chapter are twofold. On one hand, the hysteresis effect of cell puncture mechanism is analyzed, the hardware-in-the-loop simulation is performed to establish the open-loop control system of the cell puncture mechanism, and the data of the input voltage of PEAs together with the output displacement of the cell puncture mechanism are collected to support model identification; on the other hand, by regarding hysteresis and disturbance into unknown terms, the hysteretic-parameter identification is avoided, and the linear and nonlinear components of dynamics are further identified using system identification toolbox of MATLAB toolbox and PSO algorithm, respectively. The results of this chapter lay the foundation for the nonlinear robust controller designs of subsequent chapters.

REFERENCES

1. Ismail M, Ikhouane F, Rodellar J. The hysteresis Bouc–Wen model, a survey. *Archives of Computational Methods in Engineering* 2009; 16(2): 161–188.
2. Liu S, Sun D, Zhu C. Coordinated motion planning for multiple mobile robots along designed paths with formation requirement. *IEEE/ASME Transactions on Mechatronics* 2010; 16(6): 1021–1031.
3. Gu GY, Li CX, Zhu LM, et al. Modeling and identification of piezoelectric-actuated stages cascading hysteresis nonlinearity with linear dynamics. *IEEE/ASME Transactions on Mechatronics* 2015; 21(3): 1792–1797.

4 Position Tracking of Cell Puncture Mechanism Using Composite Proportional Integral Sliding Mode Control with Feedforward Control

4.1 INTRODUCTION

Cell puncture is a type of microbiological surgical technique for gene manipulation of cells in biomicroscopy [1]. Cell puncture has been favored by medical scientists, given its irreplaceable and superior role in physiology, pathology, and pharmacology [2, 3]. For example, intracytoplasmic sperm injection [4] has been widely used in assisting human reproduction by cell micromanipulation. Cell injection into zebrafish embryos has become an important method for tracking the signal transmission trajectories of human diseases and high-throughput drug analysis in vivo due to the genetic advantages of zebrafish embryos and their highly conserved disease-transmission trajectories [5]. Cell puncture is the precondition of cell injection. Cell puncture refers to the process of penetrating cell membrane and related biological tissues by injecting pipette less than 5 μm in diameter. Therefore, cell puncture is a significant technique in cell surgery. Moreover, existing studies have been conducted on adjusting the location and orientation of cells, such as optical tweezers [6, 7], electric field [8], magnetic field [9], dielectrophoresis [10], MEMS [11], ultrasound field [12], and cell puncture.

Positioning, as the key mechanism to achieve cell puncture, needs to be provided in micron or submicron precision [13]. A piezoelectric actuator (PEA) is a typical representative of intelligent materials and has been widely used in micromanipulation fields, such as cell micromanipulation and scanning probe microscopy, given its high displacement resolution, fast response, and nonpolluting property. However, PEAs are characterized by hysteresis, creep, and high-frequency vibration that can seriously reduce the accuracy of micromanipulation [14]. Thus, appropriate control strategies for position regulation of cell puncture mechanism should be adopted to compensate

DOI: 10.1201/9781003294030-4

for motion errors caused by hysteresis and ensure the rapid response and accuracy of the mechanism. Control strategies can generally be divided into open-loop and closed-loop controls. Open-loop control requires a hysteresis model of PEAs. The model is accurate and its control precision is high when the requirement of system modeling is high [15, 16]. The commonly used hysteresis models are the Preisach [17], Maxwell [18], Duhem [19], Prandtl–Ishlinskii [20], and Bouc–Wen models [21]. Scholars have completed open-loop feedforward (FF) control based on an inverse hysteresis model [22]. The combination of FF control and feedback (FB) control can be used to compensate hysteretic nonlinearity. The main points of these methods depend on the establishment and identification of accurate hysteresis models. Given the development of control theory, precise control can be achieved by designing robust controllers and considering hysteretic nonlinearity as an unknown disturbance [23, 24]. However, estimating these uncertainties is a challenging task [25].

The performance of the controller becomes satisfactory when the certainty of a given model is good [26]. Therefore, this chapter conducts parameter identification by referring to an existing hysteresis model and particle swarm optimization (PSO) [27] algorithm. An accurate dynamic model is obtained through the FF plus FB composite control strategy, wherein accurate positioning control of the cell puncture mechanism is achieved. For FB control, scholars favor sliding mode control as a simple and effective robust control method, given its powerful capability of dealing with uncertain problems [28]. Thus, a sliding mode control strategy is proposed to overcome the errors in parameter identification given the model errors. Proportional-integral control is also introduced to improve the comprehensive performance of the controller, which is applied to the experiment to improve the robustness and adaptive capability of the system.

The main contributions of this chapter are twofold. On one hand, a composite controller based on FF plus proportional-integral sliding mode control (PISMC) FB is designed to ultimately reduce errors and improve tracking control accuracy; on the other hand, the proposed control strategy is proven to have sufficient efficiency and can be easily applied to cell puncture.

The rest of this chapter is arranged as follows. Section 4.2 presents the experimental system for a cell puncture mechanism together with system dynamic model. Three kinds of controllers are described in Section 4.3. A comparative simulation study of the three controllers is completed and discussed in Section 4.4. FF control is then introduced in Section 4.5, and cell puncture experiments are performed. Section 4.6 concludes this chapter.

4.2 EXPERIMENTAL SETUP

An experimental setup for cell puncture is established as shown in Figure 4.1. The target computer is installed with National Instruments PCI-6229 (with a range of +5 V) data acquisition card. The acquisition card can achieve 32-channel A/D conversion and four-channel D/A conversion. As illustrated in Figure 4.2, the cell puncture mechanism comprises a bridge displacement mechanism and a PEA. The bridge-type displacement amplification mechanism is manufactured by 3D printing technology. The material is a mixture of photosensitive resin and polypropylene. The geometric

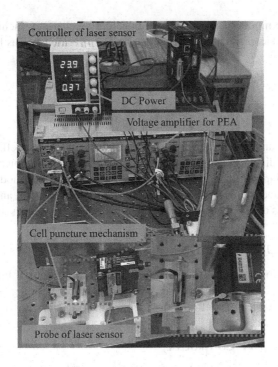

FIGURE 4.1 Global and local photographs of the experimental setup for cell puncture. (Source: S. Yu, M. Xie, H. Wu et al. /ISA Transactions 124 (2022) 427–435, with permission.)

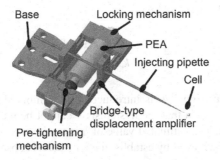

FIGURE 4.2 Mechanical architecture of the cell puncture mechanism. (Source: S. Yu, M. Xie, H. Wu et al. /ISA Transactions 124 (2022) 427–435, with permission.)

accuracy of the structure is less than 0.015 mm. PEA (model: Pst120/7/20VS12, obtained from Harbin Core Tomorrow Technology Co., Ltd.) is selected as the actuator with a displacement stroke of 20 μm. PEA output displacement is driven by a voltage amplifier (model: E00.6, obtained from Harbin Core Tomorrow Technology Co., Ltd.). The output range of the voltage amplifier is 0–120 V. The output displacement of the cell puncture mechanism is measured by a laser displacement sensor (model: LK-H020, obtained from Keyence Corporation). The measurement accuracy is 0.02 μm, whereas the range is ±3 mm.

The dynamic model of the entire cell puncture mechanism, including PEA and bridge-type displacement amplification mechanism is established as follows:

$$M\ddot{x}_d + C\dot{x} + k_e x = k_e \left(du - h \right) + f_d \tag{4.1}$$

$$\dot{h} = \zeta_1 d\dot{u} - \zeta_2 |\dot{u}| h - \zeta_3 \dot{u} |h| \tag{4.2}$$

where M, C, k_e, and x represent the equivalent mass, damping coefficient, stiffness coefficient, and displacement, respectively. In addition, d is the piezoelectric coefficient; u is the input voltage; f_d is the total disturbance of the unmodeled terms and external disturbances; and h is the hysteresis effect of the system; and ζ_1, ζ_2, and ζ_3 are the hysteresis coefficients describing the shape of the hysteresis loop.

4.3 CONTROLLER DESIGN

The goal of motion control is to allow the output displacement of the cell puncture mechanism to track the given position trajectory. The input voltage of PEA can be determined as intended in controller design once the trajectory is determined.

4.3.1 DESIGN OF FF CONTROLLER

An FF compensation controller based on the inverse Bouc–Wen model is designed in accordance with the entire dynamic model of the system. Assuming that the reference displacement is x_d and the actual displacement output is x, the expression of the actual input voltage u is as follows:

$$u = \frac{M\ddot{x}_d + C\dot{x} + k_e x}{k_e d} + \frac{h}{d} \tag{4.3}$$

The reference input signal u_d is introduced and combined with the differential equation in Eq. (4.2), given that the hysteresis h cannot be measured by sensors in practical experiments. The estimated value of h is the inverse compensation of hysteresis, which can be achieved by establishing a hysteresis observer $\hat{h}$, that is,

$$u_d = \frac{M\ddot{x}_d + C\dot{x}_d + k_e x_d}{k_e d} \tag{4.4}$$

$$\dot{\hat{h}} = \xi_1 d\dot{u}_d - \xi_2 |\dot{u}_d| \hat{h} - \xi_3 \dot{u}_d |\hat{h}| \tag{4.5}$$

Therefore, the input voltage u_{ff} of the FF controller is obtained as follows:

$$u_{ff} = \frac{M\ddot{x}_d + C\dot{x}_d + k_e x_d}{k_e d} + \frac{\hat{h}}{d} \tag{4.6}$$

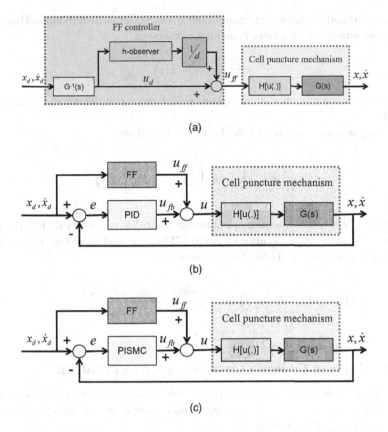

(a)

(b)

(c)

FIGURE 4.3 Flowcharts of the three controllers: (a) Principle framework of FF control based on the Bouc–Wen inverse model, (b) Composite control flowchart of FF control and PID feedback, and (c) Composite control flowchart based on FF control and PISMC feedback control. (Source: S. Yu, M. Xie, H. Wu et al. /ISA Transactions 124 (2022) 427–435, with permission.)

Figure 4.3(a) is the framework of FF control. $G^{-1}(s)$ is the inverse compensation control of vibration suppression. The inverse compensation of hysteresis is achieved by establishing a hysteresis observer.

4.3.2 DESIGN OF COMPOSITE CONTROLLER BASED ON PID FB

In practical experiments, the error of FF control often includes the following three aspects: (1) external disturbances, such as sensor noise and delay; (2) error between the established mathematical model describing the hysteresis characteristics and the actual situation; and (3) error of the hysteresis observer $\hat{h}$ in the controller and error of h in the Bouc–Wen mathematical model. Therefore, introducing an FB loop is necessary to eliminate errors. The simplest FB loop is PID FB.

The objective of the control system is to design a robust controller with an output displacement that can accurately track the reference displacement trajectory even under uncertainties, such as nonlinear hysteresis and external disturbances. Therefore, the displacement tracking error is defined as follows:

$$e(t) = x_d(t) - x(t) \tag{4.7}$$

where $x_d(t)$ is the reference displacement signal and $x(t)$ is the actual displacement signal.

The input voltage u_{fb} after FB by PID FB is as follows:

$$u_{fb}(t) = K_p e(t) + K_i \int_0^t e(\tau) d\tau + K_d \frac{d}{dt} e(t) \tag{4.8}$$

Figure 4.3(b) is the flowchart of composite controller based on PID FB. Eq. (4.6) presents the input voltage u_{ff} after FF control. The input control voltage is derived by combining equations (4.6) and (4.7) as follows:

$$u(t) = K_p e(t) + K_i \int_0^t e(\tau) d\tau + K_d \frac{de(t)}{dt} + \frac{M\ddot{x}_d + C\dot{x}_d + k_e x_d}{k_e d} + \frac{\hat{h}}{d} \tag{4.9}$$

4.3.3 DESIGN OF COMPOSITE CONTROLLER BASED ON PISMC FB

PID FB is a model-free controller. Thus, a model-based closed-loop FB controller is added to FF control, whereas the sliding mode control is used to further improve the control accuracy of the system [29].

The entire dynamic model of the system is simplified.

By taking $\bar{M} = \frac{M}{k_e d}$, $\bar{C} = \frac{C}{k_e d}$, $\tau_d = \frac{f_d - k_e h}{k_e d}$, and $D(x) = \frac{x}{d}$ and then substituting them into Eq. (4.1) yields

$$\bar{M}\ddot{x} + \bar{C}\dot{x} + D(x) = \tau - \tau_d \tag{4.10}$$

where τ_d includes the total perturbation terms of hysteretic nonlinearity and external disturbance.

Considering that errors still exist in the process of parameter identification, decomposing $\bar{M}$, $\bar{C}$, and $D(x)$ into nominal and error terms as follows:

$$\begin{cases} \bar{M} = M_0 + E_M \\ \bar{C} = C_0 + E_C \\ D(x) = D_0(x) + E_D(x) \end{cases} \tag{4.11}$$

where $M_0, C_0, D_0(x), C_0,$ and $D_0(x)$ are the nominal terms of $\bar{M}, \bar{C},$ and $D(x)$, respectively; whereas $E_M, E_C,$ and $E_D(x)$ are the parameter identification errors of $\bar{M}, \bar{C},$ and $D(x)$, respectively.

The sliding mode function is defined as follows:

$$s = \dot{e} + \Lambda e \tag{4.12}$$

where $\Lambda > 0$.

Define $\dot{x}_r = s(t) + \dot{x}(t)$, accordingly, $\ddot{x}_r = \dot{s}(t) + \ddot{x}(t)$, $\dot{x}_r = \dot{x}_d + \Lambda e$, and $\ddot{x}_r = \ddot{x}_d + \Lambda\dot{e}$. By substituting these equations into Eq. (4.8), yielding

$$\tau = M_0\ddot{x}_r + C_0\dot{x}_r + D_0(x) + E' - \bar{M}\dot{s} - \bar{C}s + \tau_d \tag{4.13}$$

where $E' = E_M\ddot{x}_r + E_C\dot{x}_r + E_D$.

Then, the controller is designed as follows:

$$\tau = \tau_m + K_ps + K_i\int_0^t s d\tau + \tau_r \tag{4.14}$$

where $K_P > 0$, $K_i > 0$. The sliding mode control also includes the proportional and integral terms. Thus, the controller is called the PISMC. The proportional term is used to amplify the control function of the deviation signal proportionally. The integral term can memorize the error, eliminate the static error, and improve system accuracy. τ_m is a control law based on a nominal model. τ_r is a robust term, which compensates for system uncertainty and disturbance. Figure 4.3(c) is the flowchart of composite controller based on PISMC FB.

$$\tau_m = M_0\ddot{x}_r + C_0\dot{x}_r + D_0(x) \tag{4.15}$$

$$\tau_r = K_r \operatorname{sgn}(s) \tag{4.16}$$

The control law is substituted into the dynamic model of the cell puncture mechanism. Specifically, Eq. (4.14) is substituted into Eq. (4.13) to analyze the stability of the PISMC FB controller. Define $E = E' + \tau_d$, and then obtain the following:

$$\bar{M}\dot{s} + \bar{C}s + K_i\int_o^t s d\tau = -K_ps - K_r\operatorname{sgn}(s) + E \tag{4.17}$$

The Lyapunov function based on integral type is designed as follows:

$$V = \frac{1}{2}s^T\bar{M}s + \frac{1}{2}\left(\int_o^t s d\tau\right)^T K_i\left(\int_o^t s d\tau\right) \tag{4.18}$$

Subsequently,

$$\dot{V} = s^T\left[\bar{M}\dot{s} + \frac{1}{2}\dot{\bar{M}}s + K_i\int_o^t s d\tau\right] \tag{4.19}$$

$s^T \left(\bar{M} - 2C \right) s = 0$ is obtained from reference [30]. Accordingly, the following equation is obtained:

$$\dot{V} = s^T \left[\bar{M}\dot{s} + \bar{C}s + K_i \int_o^t s d\tau \right] \tag{4.20}$$

By substituting Eq. (4.17) into Eq. (4.20) yields

$$\dot{V} = -s^T K_P s - s^T K_r \, \text{sgn}(s) + s^T E \tag{4.21}$$

since $K_r \geq |E|$, it is concluded that

$$\dot{V} \leq -s^T K_p s \leq 0 \tag{4.22}$$

As seen from Eq. (4.22), $s \equiv 0 \, if \, \dot{V} \equiv 0$. In conformity with LaSalle's invariance theory, $s \to 0 \, if \, t \to \infty$, that is, $e \to 0$ and $\dot{e} \to 0$.

In addition, the robustness term is designed as $\tau_r = K_r \, sat(s)$ to improve the smoothness of the control law output and eliminate buffering.

4.4   COMPARATIVE SIMULATION STUDIES

In this section, the three control schemes derived from the previous section are applied to the identified Bouc–Wen model to compensate for the hysteresis effect. Moreover, the performance indicators of different control schemes in trajectory control are compared. The definitions of the performance indicators are shown in the Appendix.

4.4.1 TRAJECTORY TRACKING SIMULATIONS

Most of the controllers, including those proposed in this chapter, are designed to track continuous and differentiable trajectories. Tracking continuous but nondifferentiable triangular wave trajectories is challenging. Therefore, the controller is used to track three kinds of trajectory curves to thoroughly verify the controller's trajectory tracking performance. The first trajectory curve is the sinusoidal signal with a frequency of 1 Hz and amplitude of 120 μm. The second trajectory curve is the sinusoidal signal whose frequency and amplitude change continuously, and its expression is as follows:

$$y = \left(65 \times \sin\left(8 \times 2 \times \pi \times e^{-0.5t} \times t - 0.5\pi \right) + 65 \right) \times 10^{-6} \times e^{-0.3t} \tag{4.23}$$

The third trajectory curve is the triangular wave signal whose amplitude and frequency constantly change.

Trajectory tracking experiments utilizing PID FB and PISMC FB controllers were conducted, and the comparative performance indicators are shown in Table 4.1.

TABLE 4.1

Performance Indicators of Different Controllers

Reference signals	Performance Indicators	FF	PID FB	PISMC FB	FF+PID FB	FF+PISMC FB
				Controller		
Sinusoidal signal	$e_{rmse}(\mu m)$	4.0184	0.2536	0.0529	0.1879	0.0272
	$e_m(\mu m)$	8.24	0.79	0.27	0.72	0.08
	$\theta(\%)$	3.89	0.66	0.23	0.60	0.08
Sinusoidal signal	$e_{rmse}(\mu m)$	4.6004	0.2397	0.0526	0.1866	0.0235
with variable	$e_m(\mu m)$	8.89	0.90	0.30	0.65	0.12
frequency and amplitude	$\theta(\%)$	5.46	0.78	0.27	0.58	0.10
Triangular wave	$e_{rmse}(\mu m)$	3.2102	0.2292	0.0514	0.1872	0.0229
trajectory	$e_m(\mu m)$	6.80	0.74	0.37	1.03	0.36
	$\theta(\%)$	3.23	0.61	0.33	0.85	0.30

Source: S. Yu, M. Xie, H. Wu et al. /ISA Transactions 124 (2022) 427–435, with permission.

The integral term plays a key role in improving the accuracy of the system. Integral term estimates and compensates for the disturbances that slowly change outside. K_r is the coefficient of robustness, wherein an appropriate K_r can stabilize the trajectory of the system on the sliding surface. However, excessive K_r will produce strong chattering, which will have a negative impact on the system. Therefore, PISMC has higher control precision than PID.

Adding FF control based on PID FB or PISMC FB improves control accuracy. The comprehensive analysis shows that FF plus PISMC FB control yields better performance than other controllers.

4.4.2 ANTI-INTERFERENCE SIMULATION

In the actual use process, the system is inevitably subjected to various external interferences. Shock interference is added with a distance of 2 m and strength of 60 V to the control system to test its anti-interference ability. The experimental results are shown in Figure 4.4. The FF controller loses its stability instantaneously when disturbed, whereas the composite control remains stable. The stabilization time of the FF plus PID FB controller is approximately 100 ms, which produces evident chattering, whereas that of the FF plus PISMC FB controller is 70 ms, which indicates minimal chattering. Therefore, the FF plus PISMC FB controller has higher control accuracy, shorter time to achieve stability, smaller convergence error, and higher robustness after being disturbed by shocks than the FF plus PID FB controller.

Increasing K_p can accelerate the convergence speed when the system is far from the sliding surface. However, excessive K_p can cause driver saturation and overshoot and make the system unstable. The integral term plays a key role in improving the accuracy of the system. K_i estimates and compensates for the disturbances that slowly

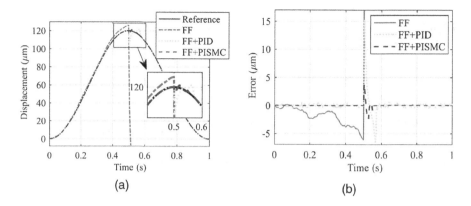

FIGURE 4.4 Tracking response under external impact interference: (a) displacement curve and (b) displacement error curve. (Source: S. Yu, M. Xie, H. Wu et al. /ISA Transactions 124 (2022) 427–435, with permission.)

change outside. Therefore, the selection of parameters K_p and K_i trades off the dynamic response and robustness of the system.

In summary, the FF plus PISMC FB controller has the advantages of FF control and sliding mode control. The FF plus PISMC FB controller achieves good steady-state response time and no evident oscillation. The FF plus PISMC FB controller considers the mathematical model and achieves better robustness compared with the FF plus PID FB controller. Therefore, the FF plus PISMC FB controller can effectively compensate and suppress the hysteresis effect of PEA and achieve precise trajectory tracking of the cell puncture mechanism. For continuous differentiable or nondifferentiable trajectory signals, the composite control can achieve accurate positioning, whereas the trajectory tracking accuracy can reach submicron levels. The FF plus PISMC FB controller has strong anti-interference performance and rapid response speed. These properties are suitable for the cell puncture mechanism.

4.5 CELL PUNCTURE EXPERIMENT

In this chapter, zebrafish embryos are punctured to verify the performance of the cell puncture mechanism. Zebrafish has been widely used in biomedical and pharmaceutical research [5] due to the 87% similarity between its genes and human genes. Zebrafish embryos can grow into small fish within 24 h, thereby allowing biologists to conduct different experiments on the same generation of fish and then study the pathological evolution process to identify a cause. Zebrafish embryos have become an important research object in life science. Transparent zebrafish embryos are also suitable for studying the cell puncture process.

An experimental device was constructed. The inverted biomicroscope (model: BA1000, obtained from Chongqing Optical Instrument Factory) was equipped with an electronic microscope (model: ZY-HDMI2800, resolution: 30 million pixels, obtained from Shenzhen Zongyuan Weiye Technology Co., Ltd.) to observe the puncture process. Injecting pipette (model: B150-86-7.5, obtained from Sutter Instrument Company) was prepared by a pipette puller (model: PC100, obtained from Narishige International Ltd.) with an end diameter of 5~μm. The holding pipette was used to absorb and fix the cells. The end of the holding pipette was smoothened and streamlined to avoid scratching the zebrafish embryos.

We connected the trajectory curve of the puncture by a sinusoidsal curve and straight line to avoid impact during the puncture process. The trajectory curve and displacement error curve of the cell puncture are shown in Figure 4.5. The process of the micropuncture of the zebrafish embryos is shown in Figure 4.6.

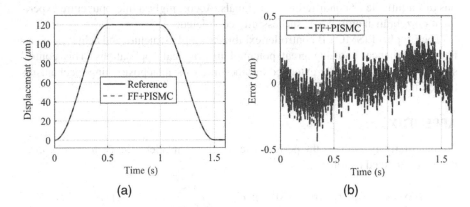

FIGURE 4.5 FF plus PISMC used for cell puncture: (a) output displacement diagram of FF plus PISMC FB controller and (b) displacement error curve of FF plus PISMC FB controller. (Source: S. Yu, M. Xie, H. Wu et al. /ISA Transactions 124 (2022) 427–435, with permission.)

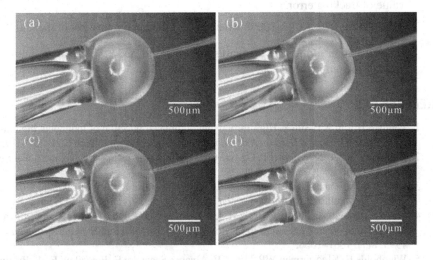

FIGURE 4.6 Micropuncture process of zebrafish embryos: (a) the cell membrane does not deform when the injecting pipette only touches the cell membrane; (b) the injecting pipette is driven by the cell puncture mechanism to squeeze the cell membrane, causing the elastic deformation of the cell membrane; (c) the injecting pipette penetrates the cell membrane, wherein the cell membrane returns to its spherical shape; and (d) the injecting pipette detaches from the cell membrane, and the cell fluid does not overflow. (Source: S. Yu, M. Xie, H. Wu et al. /ISA Transactions 124 (2022) 427–435, with permission.)

4.6 CONCLUSION

This chapter focuses on the hysteresis compensation of the cell puncture mechanism driven by a PEA to achieve high-precision motion control. The experimental results show that the compensation effect on hysteretic nonlinearity of the composite control composed of FF plus PID FB control or FF plus PISMC FB control is better than that of FF plus PID FB or FF plus PISMC FB control alone. In addition, the FF plus PISMC FB controller achieves good accuracy and robustness when tracking continuous differentiable or nondifferentiable signals. Accordingly, a micropuncture experiment on zebrafish embryos is successfully completed.

The FF plus PISMC FB controller exhibits a simple structure and achieves excellent performance, showing great potential in practical applications. This control strategy can also be easily extended to other micro-or nanopositioning mechanisms based on PEAs.

APPENDIX

To quantitatively evaluate the performance of the controllers, the performance indicators are defined as follows:

1) Root mean square error (RMSE): $e_{rmse} = \sqrt{\dfrac{1}{N}\sum\limits_{i=1}^{N}\left(x_d(i)-x(i)\right)^2}$, where N is the total number of samples. RMSE reflects the average effect of output signal tracking reference signal.

2) Maximum error: $e_m = \max |x_d(i) - x(i)|$, that is, the maximum absolute value of tracking error.

3) Nonlinearity: $\theta = \dfrac{\Delta_{\max}}{y_{FS}} \times 100\%$, where $\Delta_{\max}$ is the maximum error of the output signal after fitting with the least square method, and y_{FS} is the maximum error of the output signal.

REFERENCES

1. Wang WH, Liu XY, Sun Y. High-throughput automated injection of individual biological cells. *IEEE Transactions on Automation Science Engineering* 2009; 6(2): 209–219.
2. Braude P, Pickering S, Flinter F, Ogilvie CM. Preimplantation genetic diagnosis. *Nature Reviews Genetics* 2003; 3(12): 941–953.
3. Xie M, Li X, Wang Y, Liu Y. Sun D. Saturated PID control for the optical manipulation of biological cell. *IEEE Transactions on Control Systems Technology* 2017; 99: 1–8.
4. Van Steirteghem AC, Nagy Z, Joris H, Liu J, Staessen C, Smitz J, et al. High fertilization and implantation rates after intracytoplasmic sperm injection. *Human Reproduction* 1993; 8(7): 1061–1066.
5. Wienholds E, Kloosterman WP, Miska E, Alvarez-Saavedra E, Berezikov E, de Bruijn E. et al. MicroRNA expression in zebrafish embryonic development. *Science* 2005; 309(5732): 310–311.
6. Xie M, Shakoor A, Shen Y, Mills JK, Sun D. Out-of-plane rotation control of biological cells with a robot-tweezers manipulation system for orientation-based cell surgery. *IEEE Transactions on Biomedical Engineering* 2018; 66(1): 199–207.

7. Xie M, Shakoor A, Li C, Sun D. Robust orientation control of multi-DOF cell based on uncertainty and disturbance estimation. *International Journal of Robust and Nonlinear Control* 2019; 09(14): 4859–4871.

8. Chen J, Abdelgawad M, Yu L, Shakiba N, Chien WY, Lu Z, et al. Electrodeformation for single cell mechanical characterization. *Journal of Micromechanics and Microengineering* 2011; 21(21): 242–54012.

9. Huang CY, Hsieh TF, Chang WC, Yeh KC, Hsu MS, Chang CR, et al. Magnetic micro/nano structures for biological manipulation. *Spin* 2016; 06(01): 1650005.

10. Shafiee H, Caldwell JL, Sano MB, Davalos RV. Contactless dielectrophoresis: a new technique for cell manipulation. *Biomedical Microdevices* 2009; 11(5): 997.

11. Shakoor A, Xie M, Luo T, Hou J, Shen Y, Mills JK, et al. Achieve automated organelle biopsy on small single cells using a cell surgery robotic system. *IEEE Transactions on Biomedical Engineering* 2019; 66(8): 2210–2222.

12. Kim DH, Haake A, Sun Y, Neild AP, Ihm JE, Dual J, et al. High-throughput cell manipulation using ultrasound fields. In: *Conference Proceedings: Annual International Conference of the IEEE Engineering in Medicine and Biology Society. Conference*, vol. 4, 2004, pp. 2571–2574.

13. Huang HB, Sun D, Mills JK, Cheng SH. Robotic cell injection system with position and force control: toward automatic batch biomanipulation. *IEEE Transactions on Robotics* 2009; 25(3): 727–737.

14. Ronkanen P, Kallio P, Vilkko M, and Koivo HN. Displacement control of piezoelectric actuators using current and voltage. *IEEE/ASME Transactions on Mechatronics* 2011; 16(1): 160–166.

15. Liu P, Yan P, Zhang Z, Leng T. Modeling and control of a novel X-Y parallel piezoelectric-actuator driven nanopositioner. *ISA Transactions* 2015; 56: 145–154.

16. Agrawal A, Sun Y, Barnwell J, Raskar R. Vision-guided robot system for picking objects by casting shadows. *International Journal of Robotics Research* 2010; 29(2–3): 155–173.

17. Wei YD, Tao HF. Study the preisach model of hysteresis in piezoelectric actuator. *Piezoelectrics Acoustooptics* 2004; 26(5): 364–367.

18. Liu Y, Liu H, Wu H, Liu H. Modelling and compensation of hysteresis in piezoelectric actuators based on Maxwell approach. *Electronics Letters* 2016; 52(3): 188–190.

19. Oh J, Bernstein DS. Semilinear duhem model for rate-independent and rate-dependent hysteresis. *IEEE Transactions on Automatic Control* 2005; 50(5): 631–645.

20. Kuhnen K. Modeling, identification and compensation of complex hysteretic nonlinearities: a modified prandtl-ishlinskii approach. *European Journal of Control* 2003; 9(4): 407–418.

21. Ismail M, Ikhouane F, Rodellar J. The hysteresis Bouc–Wen model, a survey. *Archives of Computational Methods in Engineering* 2009; 16(2): 161–188.

22. Ru C, Sun L. Improving positioning accuracy of piezoelectric actuators by FF hysteresis compensation based on a new mathematical model. *Review of Scientific Instruments* 2005; 76(9): 469.

23. Chen J, Sun D, Yang J. Leader-follower formation control of multiple nonholonomic mobile robots incorporating a receding-horizon scheme. *International Journal of Robotics Research* 2010; 29(6): 727–747.

24. Kang S, Wu H, Yang X, Li Y, Wang Y. Model-free robust finite-time force tracking control for piezoelectric actuators using time-delay estimation with adaptive fuzzy compensator. *Transactions of the Institute of Measurement Control* 2020; 42(3): 351–364.

25. Xiao S, Li Y. Dynamic compensation and H∞ control for piezoelectric actuators based on the inverse Bouc–Wen model 2014; 30(1): 47–54.

26. Wu H, Sun D, Zhou Z. Model identification of a micro air vehicle in loitering flight based on attitude performance evaluation. *IEEE Transactions on Robotics* 2004; 20(4): 702–712.

27. Banks A, Vincent J, Anyakoha C. A review of particle swarm optimization. Part I: background and development. *Natural Computing* 2007; 6(4): 467–484.

28. Yu S, Ma J, Wu H, Kang S. Robust precision motion control of piezoelectric actuators using fast nonsingular terminal sliding mode with time delay estimation. *Measurement Control* 2019; 52(1–2): 11–19.

29. Hu J, Zhang P, Kao Y, Liu H, Chen D. Sliding mode control for Markovian jump repeated scalar nonlinear systems with packet dropouts: the uncertain occurrence probabi. *Applied Mathematics and Computation* 2019; 11(4): 58–79.

30. Ott C. *Cartesian Impedance Control of Redundant and Flexible-Jointrobots.* Berlin, Germany: Springer; 2008, p. 32.

5 Motion Control of Cell Puncture Mechanism Based on Fractional Non-singular Terminal Sliding Mode

5.1 INTRODUCTION

Microinjection is a micromanipulation technology that can complete biological operations in cells or embryos [1]. Specifically, microinjection can be used to inject micrometric external substances, such as drugs, sperm, DNA, protein, and RNA, into living cells [2]. Biologists can complete a series of medical research, such as in gene engineering, virus detection [3], drug development, and disease analysis [4], by observing the growth and development of cells [5]. Accordingly, microinjection has attracted the attention of many scholars [6], which has resulted in the development of many cell manipulation methods [7], such as cell localization, puncture, and injection [8]. Among these methods, cell puncture is a prerequisite of cell microinjection that uses an injecting pipette to penetrate the cell membrane and accurately locate in the designated cell tissue [9]. However, cells are small [10], easy to deform, and fragile in vitality [11]. The specific physiological characteristics of cells also require high technical requirements for cell puncture [12]. Therefore, a robust controller must be developed for a cell puncture mechanism (CPM) to achieve a sufficient precise motion control.

CPM is a mechanical and electrical integration equipment that can realize submicron-level high-precision micro/nano-operations in a sterile environment. As a newly developed intelligent driving material, piezoelectric ceramics (PEAs) [13] have no gap, no electromagnetic pollution, no noise pollution, high resolution, and large driving force [14]. Compared with other traditional hydraulic or motor drive systems, PEAs have unparalleled advantages in the field of micro-operation. However, PEAs also have obvious defects, including their limited motion range (usually within 30 microns), substantial hysteresis, creep, high-frequency vibration, and other nonlinear effects [15]. A displacement amplification mechanism must then be designed to amplify the output displacement of PEAs [16]. However, the displacement amplification mechanism can also amplify the hysteretic nonlinear effect. Therefore, the development of a high-precision, robust controller suitable for CPM presents a challenge.

DOI: 10.1201/9781003294030-5

A charge-driven control method based on constant current and high impedance can eliminate the hysteretic nonlinear effect of PEAs [17]. However, the charge driving circuit is highly complex and reduces the effective stroke of PEAs [18]. Many scholars have studied dynamic models of PEAs with hysteretic nonlinear effects [19–23]. These models provide strong theoretical support for the design of robust controllers yet often use nonlinear equations with complex structures and numerous parameters to depict the hysteretic nonlinear effect. Identifying the parameters of nonlinear equations is also a tedious and time-consuming process [24]. Therefore, hysteretic nonlinearity and external disturbance are considered unknown terms that can simplify the dynamic model to facilitate the engineering application and design of robust controllers. In robust control, sliding mode control (SMC) [25] can effectively handle the model defects and uncertainties in a nonlinear system [26], thereby making this technology suitable for the motion control of CPM.

In SMC, the displacement error and its derivatives are taken as the current state of the system [27]. The structure of the system continuously changes along with the current state and moves according to the predetermined "sliding mode" state trajectory [28]. Traditional SMC uses the linear sliding surface [29], which ensures that the state trajectory achieves progressive convergence [30]. Finite-time stability can significantly improve the performance of the controller and achieve finite-time convergence [31]. A nonsingular TSM (NTSM) [32] has been proposed to overcome the singular problem, and a fast-TSM-type reaching law has been developed to suppress chattering. Fast NTSM (FNTSM), which is achieved by synthesizing NTSM manifold and fast-TSM type reaching law, has continuous output, no chattering, and precise control [33]. However, the existing FNTSM is mostly limited to using an integer-order (IO) differentiator or integrator, while the fractional-order (FO) controller has been recently proven to outperform the IO controller [34]. Therefore, the performance of the FNTSM controller can be improved by synthesizing FO theory and FNTSM controller. Prior knowledge of the unknown quantity boundary of the system, whether by using FNTSM or FONTSM, is also required due to the unknown terms in the dynamic model of CPM. Moreover, the selection of the gain coefficient of the robust term in the controller tends to be conservative, which damages the quality of the controller. Therefore, an accurate estimation of unknown terms must be ensured.

Time-delay estimation (TDE) [35] estimates the current system state in real time from the previous system state in a closed-loop control to realize online estimation and real-time compensation for unknown items in time-delay control (TDC) [36]. TDE technology has been widely studied and applied because of its reduced dependence on the model and easy engineering application. Therefore, the nonlinear, robust controller designed by synthesizing FONTSM and TDE has strong advantages. TDE realizes a real-time estimation and compensation of unknown terms and reduces the gain coefficient in FONTSM, whereas FONTSM realizes an effective time convergence and improves control accuracy.

However, TDE has several disadvantages in practical applications that are mainly reflected in two aspects. First, in this technology, the value of time-delay parameter L must be minimized to improve the accuracy of estimating unknown items, but the adjustment of L is restricted by the electrical hardware system. Second, in TDE, both velocity and acceleration information must be known, but only a displacement sensor is often set up in practical applications. The velocity signal can be obtained by one differential calculation for the displacement signal, whereas the acceleration signal can be obtained by performing two differential calculations for the displacement signal. However, the measurement noise is greatly amplified in the differential calculation, thereby resulting in large errors in both velocity and acceleration signals. Despite realizing model-free control, TDC has substantial unknown items. In this case, controller quality can only depend on the realization of TDE technology. In the model-based controller, the proportion of unknown items in the system is effectively restrained, thereby suppressing the estimation error of TDE. Another advantage of this method is that the controller quality depends on the common implementation of FONTSM and TDE.

The major contributions of this chapter are summarized as follows. On one hand, FONTSM and TDE are synthesized to construct a nonlinear robust controller called FONTSM-TDE, which is the first time for the motion control of CPM; on the other hand, a cell puncture experiments were also performed on zebrafish embryo for the validation of the proposed FONTSM-TDE controller, demonstrating fast convergence, high control accuracy without chattering.

The rest of this chapter is arranged as follows. Section 5.2 introduces the dynamic model of CPM driven by PEAs. Section 5.3 presents the controller design together with stability analysis. Section 5.4 and 5.5 present simulation and hardware-in-loop simulation (HILS) experiments, respectively, which is followed by cell puncture experiment in Section 5.6. Section 5.7 concludes this chapter.

5.2   DYNAMIC MODEL OF CELL PUNCTURE MECHANISM

The equivalent dynamic model of CPM is established as a mass-spring-damping system as shown in Figure 5.1. Parameters b, k, and f_a denote the output forces of PEAs, damping coefficient, and stiffness coefficient of the system, respectively.

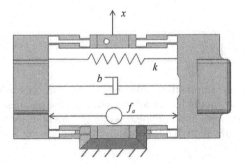

FIGURE 5.1   Equivalent dynamic model of CPM. (Source: S. Yu, H. Wu, M. Xie et al. / Precise robust motion control of cell puncture mechanism driven by piezoelectric actuators with fractional-order nonsingular terminal sliding mode control. Bio-Design and Manufacturing 3, s410–426, 2020, with permission.)

A complete dynamic model of CPM with hysteresis is constructed to comprehensively reflect the dynamic characteristics of the system by using Newton's law of motion and the Bouc–Wen model [37]. The dynamic model can be expressed as

$$m\ddot{x} + b\dot{x} + kx = f_a - \tau_d = k\left(du - h\right) - \tau_d \tag{5.1}$$

$$\dot{h} = \zeta_1 d\dot{u} - \zeta_2 |\dot{u}| h - \zeta_3 \dot{u} |h| \tag{5.2}$$

where m, d, h, and τ_d are the mass, piezoelectric coefficient, hysteresis variable, and external disturbance, respectively; x, $\dot{x}$, and $\ddot{x}$ are the actual displacement, velocity, and acceleration, respectively; and ζ_1, ζ_2, and ζ_3 are the hysteresis coefficients.

The dynamic model is then simplified reasonably, and the complexity of the model is reduced to ensure its integrity as much as possible and to facilitate the engineering application and design of the robust controller. Equation (5.1) can be rewritten as

$$\ddot{x} = \frac{kdu}{m} - \frac{kh}{m} - \frac{b}{m}\dot{x} - \frac{kx}{m} + \frac{\tau d}{m} \tag{5.3}$$

The hysteresis term and external disturbance are combined as follows into an unknown term:

$$\Delta P\left(m, k, h, \tau_d\right) = -\frac{k}{m} h + \frac{\tau_d}{m} \tag{5.4}$$

In Equations (5.2) and (5.4), the unknown term can be regarded as nonlinear terms that include hysteresis, external disturbance, and unmodeled dynamics. This unknown term has a highly complex composition, and an accurate mathematical model cannot be established. For convenience, $\Delta P(m, k, h, \tau_d)$ is abbreviated as ΔP.

Equation (5.3) is rewritten as follows to obtain a new dynamic model:

$$\ddot{x} = \frac{kd}{m} u - \frac{b}{m}\dot{x} - \frac{k}{m} x + \Delta P \tag{5.5}$$

Remark 1

All nonlinear terms are summed as unknown terms, and only a linear term exists in the optimized dynamic model [Eq. (5.5)]. Accordingly, the parameter identification efficiency is greatly improved, and nonlinear parameter identification is transformed into linear parameter identification that is conducive to engineering applications.

5.3 CONTROLLER DESIGN

5.3.1 PRIOR KNOWLEDGE

Some important knowledge of FO algorithms and TSM are introduced before presenting the controller design.

For a function $f(t)$ with respect to time t, λth-order Riemann–Liouville fractional differential and integral equations are defined as follows [38]:

$$D^{\lambda} f(t) = \frac{d^{\lambda} f(t)}{dt^{c}\lambda} = \frac{1}{\Gamma(m-\lambda)} \frac{d^{m}}{dt^{m}} \int_{t_0}^{t} \frac{f(\tau)}{(t-\tau)^{\lambda-m+1}} d\tau \tag{5.6}$$

$$_{t_0}I_t^{\lambda} f(t) = \frac{1}{\Gamma(\lambda)} \int_{t_0}^{t} \frac{f(\tau)}{(t-\tau)^{1-\lambda}} d\tau \tag{5.7}$$

$$\Gamma(x) = \int_{0}^{+\infty} t^{x-1} e^{-t} dt \tag{5.8}$$

where $m - 1 < \lambda \leq m$, $m \in N$, and $\Gamma(\cdot)$ is the Gamma function.

Lemma 1

For a Lyapunov function $V(x)$, with the any given initial value condition $V(x_0)$, the following first-order differential inequality must be satisfied [33]

$$\dot{V}(x) + \alpha V(x) + \beta V^{\gamma}(x) \leq 0 \quad 0 < \gamma < 1 \quad \alpha, \beta > 0 \tag{5.9}$$

$V(x)$ can converge to 0 in finite time and the stable time can be satisfied by the following inequality

$$T \leq \frac{1}{\alpha(1-\gamma)} \ln \frac{\alpha V^{1-\gamma}(x_0) + \beta}{\beta} \tag{5.10}$$

5.3.2 Robust Controller Design

The proposed controller aims to accurately track the desired displacement of CPM despite the presence of uncertain factors, such as external disturbance, hysteretic nonlinearity, or unmodeled items. The displacement, velocity, and acceleration errors are defined as follows:

$$\begin{cases} e = x - x_d \\ \dot{e} = \dot{x} - \dot{x}_d \\ \ddot{e} = \ddot{x} - \ddot{x}_d \end{cases} \tag{5.11}$$

where $x_d, \dot{x}_d,$ and $\ddot{x}_d$ are the desired displacement, velocity, and acceleration, respectively.

The FONTSM manifold ensures the accurate tracking and rapid response of CPM to the expected displacement and can be expressed as follows:

$$s = \dot{e} + pD^{\lambda-1}\left(\text{sig}(e)^{\alpha}\right) \tag{5.12}$$

where $0 < \alpha, \lambda < 1$, and the calculation is simplified into $\text{sig}(e)^{\alpha} = |e|^{\alpha} \, \text{sign}(e)$.

The following fast-TSM-type reaching law is adopted:

$$\dot{s} = -k_1 s - k_2 \text{sig}(s)^{\beta} \tag{5.13}$$

where $0 < \beta < 1$; and gains k_1 and k_2 are positive numbers. The controller then achieves finite time convergence.

Based on the selected FONTSM manifold s and fast-TSM-type reaching law, the designed FONTSM controller can be expressed as

$$u = \underbrace{\frac{1}{kd}\left\{ m\left[\ddot{x}_d - pD^{\lambda}\left(\text{sig}(e)^{\alpha}\right)\right] + b\dot{x} + kx - m\left[k_1 s + k_2 \text{sig}(s)^{\beta}\right]\right\}}_{\text{FONTSM}} - \underbrace{\frac{m}{kd}\Delta\hat{P}}_{\text{TDE}} \tag{5.14}$$

where $\Delta\hat{P}$ is the estimation of ΔP. In traditional methods, the online estimation of $\Delta\hat{P}$ is extremely tedious. Intelligent control methods, such as adaptive control or neural network approximation, are generally used. However, the calculation is highly complex, which is not conducive to engineering applications. Nevertheless, TDE presents a convenient method for calculating $\Delta\hat{P}$. $\Delta\hat{P}$, which is formulated as

$$\Delta\hat{P} = \Delta P(t - L) = \ddot{x}_{(t-L)} - \frac{kd}{m}u_{(t-L)} + \frac{b}{m}\dot{x}_{(t-L)} + \frac{k}{m}x_{(t-L)} \tag{5.15}$$

where t represents the current time, and L represents is the time-delay parameter. $\Delta P(t - L)$ represents value of ΔP at time $t - L$.

Reducing the time-delay parameter will improve the estimation accuracy of TDE such that $\Delta P(t) \approx \Delta P(t - L)$. The time-delay parameter can be set as being several times of the step size. The TDE error is defined as $\Delta\tilde{P} = \Delta P - \Delta\hat{P}$.

Remark 2

According to Eq. (5.15), the calculation of $\Delta\hat{P}$ is highly suitable for computer implementation.

Remark 3

Eq. (5.15) reveals that TDE is suitable for the control system without a sudden signal (e.g., impact force, friction, or clearance). The cells are soft, and elastic deformation occurs during puncture, thereby avoiding the impact force. In the mechanical structural design of CPM, a flexible hinge is used as a displacement amplification mechanism to completely prevent friction and clearance and effectively suppresses the TDE error. Therefore, TDE offers great advantages.

According to the simplified dynamic model [Eq. (5.5)] and combining Eqs. (5.14) and (5.15), the control law of the synthetic FONTSM control and TDE technology is formulated as

$$
u = \underbrace{\frac{1}{kd}\left\{ m\left[\ddot{x}_d - pD^\lambda\left(\mathrm{sig}(e)^\alpha \right) \right] + b\dot{x} + kx - m\left[k_1 s + k_2\mathrm{sig}(s)^\beta \right] \right\}}_{\text{FONTSM}}
$$

$$
\underbrace{-\frac{m}{kd}\left\{ \ddot{x}_{(t-L)} - \frac{kd}{m}u_{(t-L)} + \frac{b}{m}\dot{x}_{(t-L)} + \frac{k}{m}x_{(t-L)} \right\}}_{\text{TDE}}
\tag{5.16}
$$

By mainly including the FONTSM and TDE terms, the controller can be called FONTSM-TDE.

Remark 4

The proposed controller adopts the continuous and differentiable FONTSM manifold and the fast-TSM-type reaching law to achieve continuous output, prevent chattering, and achieve finite-time convergence. The unknown item is accurately estimated by using TDE and compensated effectively. This technology can effectively reduce the gain parameters of FONTSM and ensure the stability and smoothness of the control law output. The TDE error is compensated by the FONTSM term. Therefore, the FONTSM and TDE items perfectly complement each other.

By substituting the control law [Eq. (5.16)] into dynamic Eq. (5.5), yielding the following closed-loop system dynamic equation:

$$
\ddot{e} + pD^\lambda\left(\mathrm{sig}(e)^\alpha \right) + k_1 s + k_2\mathrm{sig}(s)^\beta = \Delta\tilde{P}
\tag{5.17}
$$

Remark 5

The solution of displacement error e is not related to the desired displacement or external disturbance, thereby validating the strong robustness of the controller.

Refer to "**Appendix A**" and "**B**" for stability analysis and parameter tuning of the closed-loop system of the controller.

5.3.3 ROBUST EXACT DIFFERENTIATOR

A laser displacement sensor is proposed in this chapter to collect the actual displacement signal. The closed-loop feedback of actual speed and actual acceleration is also required in the designed robust controller (5.16). Theoretically, the real displacement signal can be differentiated once or twice to obtain the actual velocity or acceleration, respectively. However, in practical applications, a differential operation of the displacement signal can amplify errors or result in differential explosion due to the noise signal that accompanies the displacement signal. The amplification of the noise signal in differential calculation can be restrained, and using the robust precise differentiator (RED) [39] to estimate the full state of the system can obtain results with satisfactory accuracy. In this chapter, the full-state estimation is achieved by using RED.

The second-order differentiator of RDE is expressed as

$$
\begin{aligned}
\dot{z}_0 &= v_0 = -\lambda_1 \left| z_0 - x \right|^{2/3} sign\left(z_0 - x \right) + z_1 \\
\dot{z}_1 &= v_1 = -\lambda_2 \left| z_1 - v_0 \right|^{1/2} sign\left(z_1 - v_0 \right) + z_2 \\
\dot{z}_2 &= -\lambda_3 \, sign\left(z_2 - v_1 \right)
\end{aligned}
\tag{5.18}
$$

where $\lambda_1 = 3\lambda^{1/3}$, $\lambda_2 = 1.5\lambda^{1/2}$, $\lambda_3 = 1.2\lambda$, $\lambda \geq |x|$. $z_0 = \hat{x}$, $z_1 = \hat{\dot{x}}$, and $z_2 = \hat{\ddot{x}}$ are the estimated values of RED for the displacement, velocity, and acceleration signals, respectively.

Given its direct relationship with the control accuracy of the proposed controller, the performance of RED is tested by conducting a computer simulation experiment. With the desired displacement signal, the performance of RED can be formulated as

$$
x = e^{-0.5t} \sin\left(2\pi t \right) \mu m
\tag{5.19}
$$

Meanwhile, the desired values of speed signal v and acceleration signal a are expressed as

$$
\begin{aligned}
v = \dot{x} &= -0.5e^{-0.5t} \sin\left(2\pi t \right) + 2\pi e^{-0.5t} \cos\left(2\pi t \right) \mu m \\
a = \ddot{x} &= \left[0.25e^{-0.5t} \sin\left(2\pi t \right) - \pi e^{-0.5t} \cos\left(2\pi t \right) n \right. \\
&\quad \left. - \pi e^{-0.5t} \cos\left(2\pi t \right) - 4\pi^2 e^{-0.5t} \sin\left(2\pi t \right) \right] \mu m
\end{aligned}
\tag{5.20}
$$

To accurately simulate the real signal, the Gaussian white noise signal is added as the input signal to the expected displacement. A larger λ corresponds to a higher RED estimation accuracy and a more serious chattering, whereas a smaller λ corresponds to a lower estimation accuracy and slower convergence speed. $\lambda = 0.005$ is determined after tradeoff and comparison. Figure 5.2 shows the RED test results. Despite rapidly estimating the displacement signal with the highest accuracy, RED has a poor acceleration signal estimation accuracy that directly increases the TDE error. However, FONTSM compensates for the defects in TDE precision and guarantees control precision. This viewpoint is verified by the results of the computer simulation and HILS experiments presented in Sections 5.4 and 5.5, respectively.

Based on the designed control law [Eq. (5.16)], RED is used to realize a real-time estimation of the full state. Figure 5.3 presents the control system block diagram.

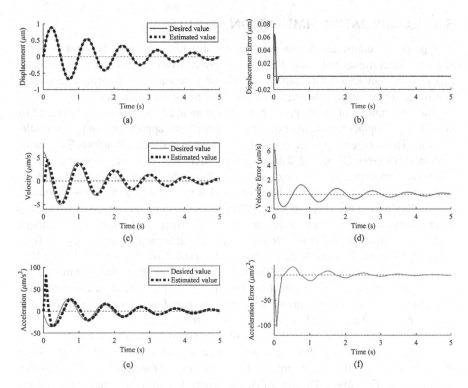

FIGURE 5.2 Performance test of RED by computer simulation. Desired and estimated values of (a) displacement, (c) velocity, and (e) acceleration. (b), (d), and (f) are the estimated error of displacement, velocity, and acceleration, respectively. (Source: S. Yu, H. Wu, M. Xie et al. / Precise robust motion control of cell puncture mechanism driven by piezoelectric actuators with fractional-order nonsingular terminal sliding mode control. Bio-Design and Manufacturing 3, 410–426, 2020, with permission.)

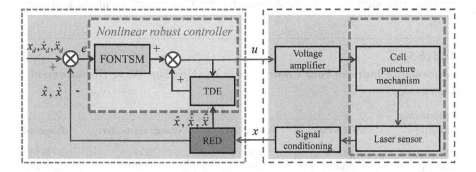

FIGURE 5.3 Controller block diagram of synthesized FONTSM, TDE, and RED technology. (Source: S. Yu, H. Wu, M. Xie et al. / Precise robust motion control of cell puncture mechanism driven by piezoelectric actuators with fractional-order nonsingular terminal sliding mode control. Bio-Design and Manufacturing 3, 410–426, 2020, with permission.)

5.4 COMPARATIVE SIMULATION STUDIES

The proposed controller has a generalized structure and an advanced concept, whereas the conventional controller can be derived from the proposed controller. In this section, computer comparative simulation experiments are performed for various controllers to investigate the performance of the proposed controller.

The performance of the proposed controller in an ideal environment is examined by conducting computer simulation experiments to verify the applicability of this controller to CPM. These experiments are conducted on a computer with a Windows 7 operating system, 3.60 GHz CPU, and 8 G memory. MATLAB/Simulink is adopted as the simulation platform with a step size of 0.1 ms. The system tracks the desired displacement curve of the sine wave, which is expressed as $y = (50 + 50 \times \sin(\pi t - \pi/2))\mu m$.

Traditional root mean square error (RMSE) and maximum error (ME) are used as evaluation indicators to quantitatively measure the performance of the four controllers.

Trial and error method is used to adjust all adjustable parameters one by one from smallest to largest. The satisfying parameters are $\alpha = 0.7$, $\beta = 0.9$, $\lambda = 0.9$, $k = 4 \times 10^{-6}$, $k_1 = -2 \times 10^4$, and $k_2 = -0.05$. The parameters of the other three controllers are consistent with the above parameters. In MF-FNTSM and Wang's controller, the value of $\bar{m}$ is reduced to play a role similar to low-pass filtering to decrease the influence of noise in the control law and improve control quality.

Table 5.1 and Figure 5.4 show that all four controllers can accurately achieve sinusoidal motion. Although slightly greater than that of MB-FNTSM, the RMSE of the proposed controller is substantially lower than that of the MB-FNTSM controller in terms of ME (Table 5.1). Generally, the proposed controller obtains the highest control accuracy. The displacement error curves of MF-FNTSM and Wang's controller regularly fluctuate, and their fluctuation period is the same as that of the desired displacement. A comparative study of the TDE error, the output terms of the control law, is conducted to explain this phenomenon and to reveal the working mechanism of the proposed controller.

The TDE error curve in Figure 5.5 is drawn by the TDE error equation $\Delta \tilde{P} = \Delta P - \Delta \hat{P}$. TDE accurately estimates the unknown terms, and the TDE error of the four controllers is controlled in the order of 10^{-3}. Compared with the maximum

---

TABLE 5.1

Performance Index of Four Controllers in Tracking Sine Wave Displacement Curve

Performance Indicator	Controller			
	MF-FNTSM	MB-FNTSM	Wang's controller	Proposed
RMSE (μm)	4.91×10^{-4}	1.02×10^{-4}	1.58×10^{-4}	1.11×10^{-4}
ME (μm)	8.66×10^{-4}	5.22×10^{-4}	4.53×10^{-4}	2.36×10^{-4}

Source: S. Yu, H. Wu, M. Xie et al. / Precise robust motion control of cell puncture mechanism driven by piezoelectric actuators with fractional-order nonsingular terminal sliding mode control. Bio-Design and Manufacturing 3, 410–426, 2020, with permission.

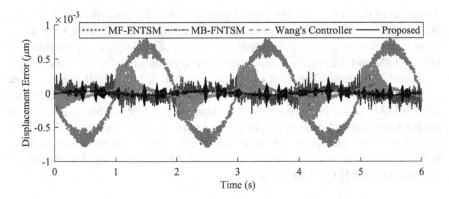

FIGURE 5.4 Displacement error curve of four controllers in tracking the sine wave displacement curve. (Source: S. Yu, H. Wu, M. Xie et al. / Precise robust motion control of cell puncture mechanism driven by piezoelectric actuators with fractional-order nonsingular terminal sliding mode control. Bio-Design and Manufacturing 3, 410–426, 2020, with permission.)

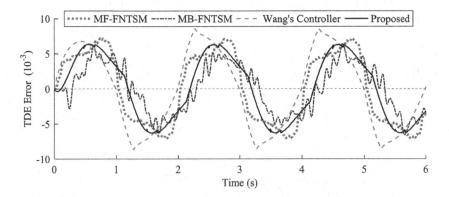

FIGURE 5.5 Estimation error curve of the unknown term in tracking the sine wave displacement: a MF-FNTSM, b MB-FNTSM, c Wang's controller, and d proposed controller. (Source: S. Yu, H. Wu, M. Xie et al. / Precise robust motion control of cell puncture mechanism driven by piezoelectric actuators with fractional-order nonsingular terminal sliding mode control. Bio-Design and Manufacturing 3, 410–426, 2020, with permission.)

output value of the control law (60 V), the TDE error is nearly ignored. Moreover, the TDE error curve and desired displacement error curve fluctuate synchronously in MF-FNTSM and Wang's controller. Therefore, the TDE error directly affects control accuracy. However, this phenomenon is not observed in MB-FNTSM and the proposed controller. Therefore, the composition of each item in the control law must be comprehensively investigated.

Figure 5.6 presents the controller outputs, including total control law (black solid line), sliding mode term (short dotted line), and TDE term (long dotted line). For the model-free controllers (i.e., MF-FNTSM and Wang's controller), the output of the sliding mode term is approximately zero, whereas that of the controller depends on the output of the TDE term. The control quality of the TDE term directly determines the greatest overall control effect because the sliding mode and TDE term do not have complementary roles. Meanwhile, the outputs of model-based controllers (i.e., MB-FNTSM and the proposed controller) depend on the combined action of sliding mode and TDE terms. The TDE error fluctuates periodically, but the sliding mode term compensates for the TDE error and achieves precise control. Therefore, MB-FNTSM and the proposed controller obtain the smallest RMSE. u_{TDE} represents the unknown term, whereas u_{FNTSM} or u_{FONTSM} represent the known term. The proportion of unknown and known items in the model verifies that the dynamic model includes substantial unknown terms, such as hysteretic nonlinearity, disturbance, and unmodeled terms [Figure 5.6 (b) and (d)].

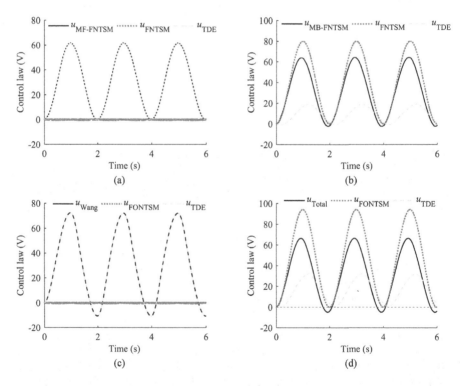

FIGURE 5.6 Output curves of the controller in computer simulation. Total control law (black solid line), sliding mode term (pink dotted line), and TDE term (blue dotted line): (a) MF-FNTSM, (b) MB-FNTSM, (c) Wang's controller, and (d) proposed controller. (Source: S. Yu, H. Wu, M. Xie et al. / Precise robust motion control of cell puncture mechanism driven by piezoelectric actuators with fractional-order nonsingular terminal sliding mode control. Bio-Design and Manufacturing 3, 410–426, 2020, with permission.)

Remark 6

In model-free control, the substantial unknown information must be estimated by using TDE, and control quality can be improved by reducing the value of $\bar{m}$. Therefore, unknown items occupy a dominant position. As a result, the output of the control law heavily depends on the realization of TDE. In engineering applications, the accuracy of TDE is restricted by sensor measurement noise and CPU operation speed, thereby increasing the time-delay parameter and further reducing the control quality. In model-based control, Dynamic Eq. (5.5) contains the physical information of the model, and the proportion of unknown terms is greatly reduced. Therefore, the sliding mode term can play its role and occupy the main position of the control law output. u_{FNTSM} or u_{FONTSM} can achieve finite-time convergence, thereby improving the control accuracy of the controller.

In model-based control, Dynamic Eq. (5.5) reflects most of the model information, and the proportion of unknown terms is greatly reduced. Therefore, the sliding mode term can play its role and occupy the main position of the control law output. The FONTSM term can achieve finite-time convergence, thereby improving the control accuracy of the controller. Specifically, the FO term in the FONTSM manifold further improves the control effect of SMC.

5.5   SEMI-PHYSICAL DISPLACEMENT TRACKING EXPERIMENT

The performance of four controllers in a real environment is compared by conducting an HILS experiment [40].

Figure 5.7 shows the HILS system constructed by xPC Target. Figure 5.8 shows the hardware system used in the cell puncture displacement tracking experiment. xPC Target provides a complete solution of rapid prototyping and HILS and can implement the control strategy of the Simulink program running on a PC to act on a physical device. The PEAs and voltage amplifiers are obtained from Harbin Core Tomorrow Science Technology Co., Ltd., and their models are Pst120/7/20VS12 and E00.6, respectively. The laser displacement sensor is obtained from KEYENCE Corporation, and its model and resolution are lk-h020 and 0.02 μm, respectively.

MATLAB/Simulink 2016b is used to build the controller on the host computer to compile and generate the target code and to download this code to the target computer. After converting the digital signal output of the target computer into an analog voltage signal, this signal is outputted to the voltage amplifier and drives the CPM to move. After the analog/digital conversion, the analog displacement signal collected by the laser displacement sensor is fed back to the target computer. The HILS system is then constructed.

The controller design must consider tracking the continuous and differentiable trajectory curve. However, a continuous yet undifferentiable trajectory curve is tracked in the HILS experiment to comprehensively examine the performance of the controller. This situation presents a challenge to the controller and can test its effectiveness in real engineering applications. Therefore, further research is necessary. Figure 5.9 shows the desired displacement curve.

The displacement error curve in Figure 5.10 and the performance index in Table 5.2 indicate that compared with the computer simulation results, the accuracy

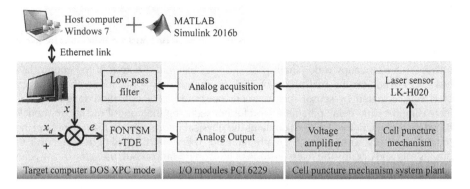

FIGURE 5.7 Construction of HILS system of CPM based on xPC Target technology. (Source: S. Yu, H. Wu, M. Xie et al. / Precise robust motion control of cell puncture mechanism driven by piezoelectric actuators with fractional-order nonsingular terminal sliding mode control. Bio-Design and Manufacturing 3, 410–426, 2020, with permission.)

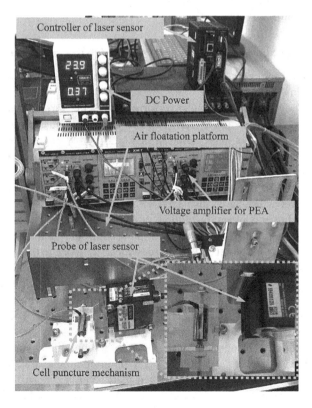

FIGURE 5.8 Global picture and local structure of displacement tracking experiment of CPM. (Source: S. Yu, H. Wu, M. Xie et al. / Precise robust motion control of cell puncture mechanism driven by piezoelectric actuators with fractional-order nonsingular terminal sliding mode control. Bio-Design and Manufacturing 3, 410–426, 2020, with permission.)

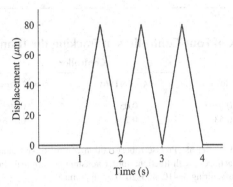

FIGURE 5.9 Triangular curve as the desired displacement in the HILS experiment. (Source: S. Yu, H. Wu, M. Xie et al. / Precise robust motion control of cell puncture mechanism driven by piezoelectric actuators with fractional-order nonsingular terminal sliding mode control. Bio-Design and Manufacturing 3, 410–426, 2020, with permission.)

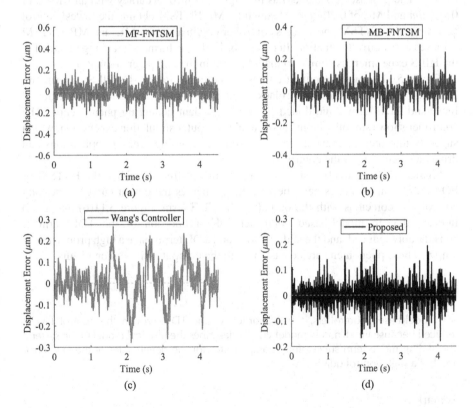

FIGURE 5.10 Displacement error curve of four controllers when tracking the triangular curve: (a) MF-FNTSM, (b) MB-FNTSM, (c) Wang's controller, and (d) proposed controller. (Source: S. Yu, H. Wu, M. Xie et al. / Precise robust motion control of cell puncture mechanism driven by piezoelectric actuators with fractional-order nonsingular terminal sliding mode control. Bio-Design and Manufacturing 3, 410–426, 2020, with permission.)

TABLE 5.2

Performance Index of Four Controllers in Tracking the Triangular Curve

Performance	Controller			
Indicator	MF-FNTSM	MB-FNTSM	Wang' controller	Proposed
RMSE (μm)	0.077	0.057	0.093	0.041
ME (μm)	0.532	0.294	0.298	0.204

Source: S. Yu, H. Wu, M. Xie et al. /Precise robust motion control of cell puncture mechanism driven by piezoelectric actuators with fractional-order nonsingular terminal sliding mode control. Bio-Design and Manufacturing 3, 410–426, 2020, with permission.

of HILS is reduced under the effects of sensor resolution, measurement noise, and external disturbance. However, the four controllers can achieve an accurate triangular motion. The proposed controller has the highest control accuracy with an RMSE of 0.041 μm and ME of 0.204 μm. Meanwhile, MF-FNTSM obtains the lowest control accuracy, MB-FNTSM and Wang's controller obtain a similar ME, and MB-FNTSM achieves a low error fluctuation. In other words, the performance ranking obtained in the HILS experiment is similar to that obtained in the computer simulation.

Figure 5.11 shows the output curves of the controller in the HILS experiment, including the total control law (black solid line), sliding mode term (pink dotted line), and TDE term (blue dotted line) with evident chattering phenomenon. The controller shows smooth output curves in the computer simulation experiment, which suggests that measurement noise and disturbance in engineering applications can drive the chattering of the control law.

Figure 5.11(a) and 11(c) show that in the model-free controller, the FNTSM or FONTSM term oscillates near the zero line, whereas the output curve of the total control law coincides with the output of the TDE term. Figure 5.11(b) and 11(d) indicate that in model-based controllers, the proportion of the TDE term is considerably reduced, and the sliding mode and TDE terms have a high proportion of output. These phenomena are consistent with the computer simulation results.

Remark 7

For model-based controllers, the proportion of the TDE term in the control law is reduced because the dynamic model vividly describes the physical model of the system. The sliding mode term successfully compensates for the deficiencies of the TDE error and has a high control quality.

Remark 8

In the HILS experiment, the output of the control law shows apparent chattering, and the accuracy control is considerably reduced due to the unfavorable constraints of sensor resolution, measurement noise, and external disturbance compared with the computer

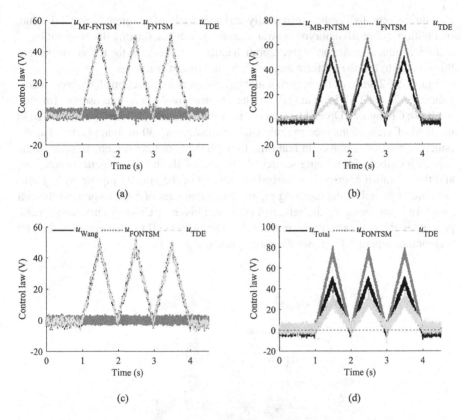

FIGURE 5.11 Output curves of the controller in the HILS experiment, including the total control law (black solid line), sliding mode term (pink dotted line), and TDE term (blue dotted line): (a) MF-FNTSM, (b) MB-FNTSM, (c) Wang's controller, and (d) proposed controller. (Source: S. Yu, H. Wu, M. Xie et al. / Precise robust motion control of cell puncture mechanism driven by piezoelectric actuators with fractional-order nonsingular terminal sliding mode control. Bio-Design and Manufacturing 3, 410–426, 2020, with permission.)

simulation experiment. However, the proposed controller retains the highest control accuracy. The proposed controller also shows high stability, strong robustness, and precise control accuracy and can drive the CPM to achieve a precise trajectory tracking control.

5.6 CELL PUNCTURE EXPERIMENT

Zebrafish and human genes have 87% familiarity [41]. Therefore, zebrafish has become the third most widely used vertebrate model organism used in life science research after mice and rats [42]. Disease research, new drug development, chemical safety, and environmental toxicology monitoring can be realized by injecting different substances into zebrafish embryos. These embryos must complete a large sample volume of microinjection within a few hours after spawning. Therefore, puncturing zebrafish embryo efficiently and accurately offers great biological value. In addition, zebrafish embryos are transparent, thereby facilitating the observation of cell micropuncture. Accordingly, zebrafish embryo is treated as the research object in this section to test the performance of CPM and robust controller.

Figure 5.12 presents the experimental environment of the zebrafish embryo micropuncture. The CPM is installed on the inverted biological microscope (model: BA1000; Chongqing Optical Instrument Factory). The zebrafish embryo is located in the field of view of the electron microscope (resolution: 30 million pixels), and the puncture process is played in real time through the computer screen. To prevent the zebrafish embryo from being scratched, the end of the holding pipette is polished, and the zebrafish embryo is adsorbed on the end of the holding pipette by negative pressure. The end of the injecting pipette has a diameter of 5 μm to prevent the cell puncture from damaging the zebrafish embryo. Driven by CPM, the injecting pipette penetrates the cell membrane, reaches the designated cell tissue, and exits the cell membrane. Figure 5.13 shows the entire puncture process.

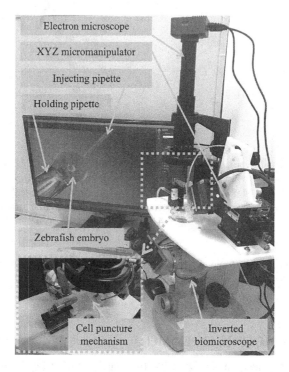

FIGURE 5.12 Zebrafish embryo micropuncture experiment under an electron microscope. (Source: S. Yu, H. Wu, M. Xie et al. / Precise robust motion control of cell puncture mechanism driven by piezoelectric actuators with fractional-order nonsingular terminal sliding mode control. Bio-Design and Manufacturing 3, 410–426, 2020, with permission.)

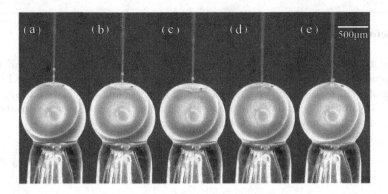

FIGURE 5.13 Electron microscopic pictures of zebrafish embryo for micropuncture. (a) Injecting the pipette effortlessly contacts the outside of the cell membrane. (b), (c) CPM continuously increases the displacement and feed, and the injecting pipette causes the cell membrane to deform substantially. (d) Head of the injecting pipette entering the zebrafish embryo and reaching the designated position. (e) Injecting pipette exiting from the zebrafish embryo. (Source: S. Yu, H. Wu, M. Xie et al. / Precise robust motion control of cell puncture mechanism driven by piezoelectric actuators with fractional-order nonsingular terminal sliding mode control. Bio-Design and Manufacturing 3, 410–426, 2020, with permission.)

5.7 CONCLUSION

This chapter designs a novel, nonlinear robust controller for the developed CPM driven by PEAs. To facilitate engineering applications, the dynamic model of CPM is locally optimized. The new controller is based on the simplified Bouc–Wen dynamic model, which includes TDE and FONTSM terms. The real-time estimation and compensation of unknown items are realized via TDE. The FONTSM term adopts the FONTSM manifold and fast-TSM-type reaching law, which can achieve finite-time convergence and high control accuracy. The stability of the developed scheme is proven using Lyapunov theory. Compared with IO or model-free controllers, the proposed controller has stronger robustness and higher control accuracy. The cell puncture experiments reveal that the proposed controller can accurately control the micropuncture mechanism to effectively micropuncture zebrafish embryos. Puncture force is introduced in the robust controller to further improve the cell survival rate. The displacement and expected force are used as tracking targets before and during the cell puncture process, respectively. Future studies should track the force and displacement signals simultaneously in the cell puncture process and realize compliance control.

APPENDIX

STABILITY ANALYSIS

For TDE error $\Delta \tilde{P}$, a positive number φ makes $\Delta \tilde{P}$ bounded, i.e., $\left| \Delta \tilde{P} \right| \le \varphi$. The boundedness certificate of $\Delta \tilde{P}$ can be referred [35, 36].

The proof of Definition 2 and the stability of the closed loop system are demonstrated by using Lyapunov theory.

Definition 2

For the CPM described in Equations (5.1) and (5.2), if the FONTSM manifold represented by Eq. (5.12) and the fast-TSM-type reaching law represented by Equation (6.13) are adopted, then the closed-loop system can achieve finite-time stability under the action of the proposed controller (Eq. (5.16)). The system-state trajectory can converge to the following range:

$$|s| \leq \min\{d_1, d_2\} \quad d_1 = \frac{\varphi}{k_1} \quad d_2 = \left(\frac{\varphi}{k_2}\right)^{1/\beta} \tag{5.21}$$

Proof

A Lyapunov function is established as follows:

$$V = \frac{1}{2}s^2 \tag{5.22}$$

The above function is differentiated with respect to time and then combined with the FONTSM manifold (6.12) to obtain

$$\dot{V} = s\left(\ddot{e} + pD^\lambda \text{sig}(e)^\alpha\right) \tag{5.23}$$

Eq. (5.17) is substituted in the Eq. (5.23) and yields

$$\dot{V} = -s\left(k_1 s + k_2 \text{sig}(s)^\beta - \Delta \tilde{P}\right) \tag{5.24}$$

Combined with boundary condition $|\Delta \tilde{P}| \leq \varphi$, the Eq. (5.24) is transformed into inequality, as follows:

$$\dot{V} \leq -s\left(k_1 s + k_2 \text{sig}(s)^\beta - \varphi\right) \tag{5.25}$$

The above inequality can be converted into the bellowing forms:

$$\dot{V} \leq -s\left[\left(k_1 - \frac{\varphi}{s}\right)s + k_2 \text{sig}(s)^\beta\right] \tag{5.26}$$

$$\dot{V} \leq -s\left[k_1 s + \left(k_2 - \frac{\varphi}{\text{sig}(s)^\beta}\right)\text{sig}(s)^\beta\right] \tag{5.27}$$

For Eq. (5.26), $|s| > \varphi/k_1$, then $\dot{V} < 0$. Thus, Eq. (5.26) can be rewritten as follows:

$$\dot{V} \le -s\left(\bar{k}_1 s + k_2 \text{sig}(s)^\beta\right) = -\bar{k}_1 s^2 - sk_2 \text{sig}(s)^\beta \qquad (5.28)$$

where $\bar{k}_1 = k_1 - \dfrac{\varphi}{s}$.

Eq. (5.28) is re-arranged as

$$\dot{V} + 2\bar{k}_1 V + 2^{\frac{1+\beta}{2}} k_2 V^{\frac{1+\beta}{2}} \le 0 \qquad (5.29)$$

According to Lemma 1, V converges to 0 in a finite time. The convergence time T_1 satisfies

$$T_1 \le \frac{1}{\bar{k}_1(1-\beta)} \ln\left(\frac{2\bar{k}_1 V_1^{\frac{1-\beta}{2}}(x_0)}{2^{\frac{1+\beta}{2}} k_2} + 1\right) \qquad (5.30)$$

In finite time T_1, the system-state trajectory s converges to the following region:

$$|s| \le \frac{\varphi}{k_1} = d_1 \qquad (5.31)$$

Using similar analysis for Eq. (5.27), if $|s|\,\beta > \varphi/k_2$, then $\dot{V} < 0$. Thus, Eq. (5.27) can be rewritten as follows:

$$\dot{V} \le -s\left(k_1 s + \bar{k}_2 \text{sig}(s)^\beta\right) = -k_1 s^2 - s\bar{k}_2 \text{sig}(s)^\beta \qquad (5.31)$$

where $\bar{k}_2 = k_2 - \dfrac{\varphi}{\text{sig}(s)^\beta}$.

Eq. (5.31) is re-arranged as follows:

$$\dot{V} + 2k_1 V + 2^{\frac{1+\beta}{2}} \bar{k}_2 V^{\frac{1+\beta}{2}} \le 0 \qquad (5.32)$$

According to Lemma 1, V converges to 0 in a finite time. The convergence time T_2 satisfies

$$T_2 \le \frac{1}{k_1(1-\beta)} \ln\left(\frac{2k_1 V_1^{\frac{1-\beta}{2}}(x_0)}{2^{\frac{1+\beta}{2}} \bar{k}_2} + 1\right) \qquad (5.33)$$

In finite time T_2, the system-state trajectory s converges to the following region:

$$|s| \leq \left(\frac{\varphi}{k_2}\right)^{\frac{1}{\beta}} = d_2 \tag{5.34}$$

In sum, the system-state trajectory converges to the following range in finite time:

$$|s| \leq \min\{d_1, d_2\} \tag{5.35}$$

Therefore, the stability of the closed-loop system is certain, and the proof of Definition 2 is completed.

CONTROLLER ADJUSTMENT

Six parameters, namely, α, β, k, k_1, k_2, and λ must be tuned in the proposed controller.

Remark 9

From Eqs (5.30), (5.33), a smaller β corresponds to a shorter convergence time of the system-state trajectory and a higher accuracy of the control system.

Remark 10

From Eqs (5.12), (5.13), and (5.16), when $\alpha \to 1$, $\beta \to 1$, the sign function disappears, and the controller shows linear characteristics. When $\alpha \to 0$, $\beta \to 0$, the control law is discontinuous under the influence of the symbolic function. Therefore, $0 < \alpha, \beta < 1$ ensures that the controller exhibits not only the chattering-free characteristics of linear control but also the strong, robust characteristics of a discontinuous controller.

Remark 11

From Eq. (5.21), the k_1 and k_2 gains of the reaching law affect the convergence region of the state trajectory. Gains k_1 and k_2 must be satisfied as $k_1 = k_2 > |\Delta \tilde{P}|$ to make the state trajectory converge to a small region as soon as possible.

Remark 12

The FO term $D^\lambda(\text{sig}(e)^\alpha)$ can enhance the sign function and show a large amplitude when the sign of displacement error changes. When $\lambda \to 1$, the amplitude increases. When $\lambda \to 0$, the effect of FO is weakened and reduced to IO. p is used to increase the variation amplitude of the FO term.

THREE CONTROLLERS FOR COMPARISON

Three existing controllers are derived from the proposed controller as prototypes in the comparative experiments. These three controllers use the FONTSM manifold, the fast-TSM-type reaching law, and the TDE technology to build an FO controller without a model and apply the NTSM manifold, the fast-TSM-type reaching law, and the TDE technology to build an IO controller whether based on a model or not.

CONTROL 1: WANG'S CONTROLLER

The model of System Dynamics (5.1) and (5.2) is simplified as

$$\bar{m}\ddot{x} + N\left(x, \dot{x}, \ddot{x}, k, d, h, \tau_d\right) = u \tag{5.36}$$

where $N\left(x, \dot{x}, \ddot{x}, k, d, h, \tau_d\right) = \left(\dfrac{m}{kd} - \bar{m}\right)\ddot{x} + \dfrac{1}{k}\dot{x} + \dfrac{1}{d}x + \dfrac{h}{d} - \dfrac{\tau_d}{kd}$. To simplify the expression, $N(t)$ represents state of $N\left(x, \dot{x}, \ddot{x}, k, d, h, \tau_d\right)$ at time t, and the simplified dynamic model is reformulated as

$$\bar{m}\ddot{x} + N\left(t\right) = u \tag{5.37}$$

Based on Dynamic Model (5.37), $N(t)$ is estimated by TDE technology and is expressed as

$$N\left(t\right) \approx \hat{N}\left(t\right) = N\left(t - L\right) = u\left(t - L\right) - \bar{m}\ddot{x}\left(t - L\right) \tag{5.38}$$

The FONTSM sliding surface and fast-TSM-type reaching law described in Eqs (5.12) and (5.13) are used to construct a model-free, robust controller based on FONTSM and TDE. The designed control law imitates the following design ideas and methods of the control law provided in

$$u = \bar{m}\left(\ddot{x}_u + pD^\lambda\left(\text{sig}\left(e\right)^\alpha\right) + k_1 s + k_2 \text{sig}\left(s\right)^\beta\right) + u_{(t-L)} - \bar{m}\ddot{x}_{(t-L)} \tag{5.39}$$

According to its proponent, this controller is called the Wang's controller.

CONTROL 2: MB-FNTSM CONTROLLER

If $\lambda = 0$, then the FO function in the controller is lost. The NTSM manifold is formulated as

$$s = \dot{e} + p\,\text{sig}\left(e\right)^\alpha \tag{5.40}$$

By using the fast-TSM-type reaching law (Eq. (5.17)) and the dynamic model (Eq. (5.5)), the controller is constructed as

$$u = \underbrace{\frac{1}{kd}\left\{ m\left[\ddot{x}_d - p\alpha\dot{e}|e|^{\alpha-1}\right] + b\dot{x} + kx - m\left(k_1 s + k_2 \text{sig}(s)^{\beta}\right)\right\}}_{\text{FNTSM}} - \underbrace{\frac{m}{kd}\Delta\hat{P}}_{\text{TDE}} \quad (5.41)$$

NTSM manifold and fast-TSM-type reaching law are adopted in the sliding mode term of the proposed controller. This term is labeled FNTSM, and the model-based controller is abbreviated to MB-FNTSM for convenience.

Substituting the control law [Eq. (5.42)] into the dynamic model [Eq. (5.5)] leads to the following displacement error equation of the closed-loop system:

$$\ddot{x} + p\alpha\dot{e}\,\text{sig}(e)^{\alpha-1} + k_1 s + k_2 \text{sig}(s)^{\beta} = \Delta\tilde{P} \quad (5.42)$$

The difference between Eqs. (5.17) and (5.42) is called the FO term. A stability analysis of the MB-FNTSM controller can also be completed by referring to the analysis of the proposed controller.

CONTROL 3:   MF-FNTSM CONTROLLER

Let $\lambda = 0$. Then, the MB-FNTSM controller can be designed.

The unknown term is estimated by Eq. (5.38) using the dynamic model [Eq. (5.37)], NTSM manifold [Eq. (5.40)], and fast-TSM-type reaching law [Eq. (5.13)]. The control law can be designed by referring to the idea presented in [28] as shown below:

$$u = \underbrace{\bar{m}\left(\ddot{x}_a + p\alpha\dot{e}|e|^{\alpha-1} + k_1 s + k_2 \text{sig}(s)^{\beta}\right)}_{\text{FNTSM}} + \underbrace{u_{(t-L)} - \bar{m}\ddot{x}_{(t-L)}}_{\text{TDE}} \quad (5.43)$$

The controller is abbreviated as MF-FNTSM.

Remark 13

The complexity of the controller reflects the required amount of calculation and debugging. The four controllers are ranked as follows in a descending order according to their required amount of calculation: proposed controller>Wang's controller>MB-FNTSM>MF-FNTSM. In engineering applications, the control accuracy and calculation amount of these controllers can be comprehensively balanced to determine the appropriate controller.

REFERENCES

1. Srikumar KCX, Zhengyi Z. Guided cell migration on a graded micropillar substrate. *Bio-Design Manufacturing* 2020; 3(1): 60–70.
2. Permana S, Grant E, Walker GM, Yoder JA. A review of automated microinjection systems for single cells in the embryogenesis stage. *IEEE-ASME Transactions on Mechatronics* 2016; 21(5): 2391–2404.
3. Lu Z, Chen PCY, Nam J, Ge R, Lin W. A micromanipulation system with dynamic force-feedback for automatic batch microinjection. *Journal of Micromechanics and Microengineering* 2007; 17(2): 314–321.
4. Liu X, Kim K, Zhang Y, Sun Y. Nanonewton force sensing and control in microrobotic cell manipulation. *The International Journal of Robotics Research* 2009; 28(8): 1065–1076.
5. Jingyu OWL A high-throughput three-dimensional cell culture platform for drug screening. *Bio-Design and Manufacturing* 2020; 3(1): 40–47.
6. Wang W, Liu X, Gelinas D, Ciruna B, Sun Y. A fully automated robotic system for microinjection of zebrafish embryos. *PLoS One* 2007; 2(9): 1.
7. Xie M, Shakoor A, Shen Y, Mills JK, Sun D. Out-of-plane rotation control of biological cells with a robot-tweezers manipulation system for orientation-based cell surgery. *IEEE Transactions on Biomedical Engineering* 2019; 66(1): 199–207.
8. Sun Y, Nelson BJ. Biological cell injection using an autonomous microrobotic system. *The International Journal of Robotics Research* 2002; 21(10): 861–868.
9. Kuchimaru T, Kataoka N, Nakagawa K, Isozaki T, Miyabara H, Minegishi M, Kadonosono T, Kizakakondoh S. A reliable murine model of bone metastasis by injecting cancer cells through caudal arteries. *Nature Communications* 2018; 9(1): 2981.
10. Liu X, Fernandes R, Gertsenstein M, Perumalsamy A, Lai I, Chi MMY, Moley KH, Greenblatt E, Jurisica I, Casper RF et al. Automated microinjection of recombinant BCL-x into mouse zygotes enhances embryo development. *PLoS One* 2011; 6(7): 1.
11. Ma L, Xu X, Jiang W. One-dimensional microstructure assisted intradermal and intracellular delivery. *Bio-Design and Manufacturing* 2019; 2(2): 24–30.
12. Ingber DE. Tensegrity I. Cell structure and hierarchical systems biology. *Journal of Cell Science* 2003; 116(7): 1157–1173.
13. Yu S, Xie M, Wu H, Ma J, Li Y, Gu H. Composite proportional-integral sliding mode control with feedforward control for cell puncture mechanism with piezoelectric actuation. *ISA Transactions* 2020. doi: 10.1016/j.isatra.2020.02.015.
14. Wang C, Lou X, Xia T, Tian S. The dielectric, strain and energy storage density of BNT-BKHXT1-x piezoelectric ceramics. *Ceramics Internationa* 2017; 43(12): 9253–9258.
15. Zhang Y, Wang S, Chen C, Zhang N, Wang A, Zhu Y, Cai F. Reduced hysteresis of KNNS-BNKZ piezoelectric ceramics through the control of sintering temperature. *Ceramics International* 2018; 44(11): 12435–12441.
16. Xu Q. Precision motion control of piezoelectric nanopositioning stage with chattering-free adaptive sliding mode control. *IEEE Transactions on Automation Science and Engineering* 2017; 14(1): 238–248.
17. Moheimani SOR, Vautier BJG. Resonant control of structural vibration using charge-driven piezoelectric actuators. *IEEE Transactions on Control Systems Technology* 2005; 13(6): 1021–1035.
18. Main JA, Garcia E, Newton DV. Precision position control of piezoelectric actuators using charge feedback. *Journal of Guidance, Control, and Dynamics* 1995; 18(5): 1068–1073.

19. Xiao S, Li Y. Modeling and high dynamic compensating the rate-dependent hysteresis of piezoelectric actuators via a novel modified inverse preisach model. *IEEE Transactions on Control Systems Technology* 2013; 21(5): 1549–1557.
20. Zhang A, Wang B. The influence of Maxwell stresses on the fracture mechanics of piezoelectric materials. *Mechanics of Materials* 2014; 68: 64–69.
21. Lin C, Lin P. Tracking control of a biaxial piezo-actuated positioning stage using generalized Duhem model. *Computers & Mathematics with Applications* 2012; 64(5): 766–787.
22. Jiang H, Ji H, Qiu J, Chen Y. A modified Prandtl-Ishlinskii model for modeling asymmetric hysteresis of piezoelectric actuators. *IEEE Transactions on Ultrasonics, Ferroelectrics, and Frequency Control* 2010; 57(5): 1200–1210.
23. Rakotondrabe M. Bouc-wen modeling and inverse multiplicative structure to compensate hysteresis nonlinearity in piezoelectric actuators. *IEEE Transactions on Automation Science and Engineering* 2011; 8(2): 428–431.
24. Wen Z, Ding Y, Liu P, Ding H. An efficient identification method for dynamic systems with coupled hysteresis and linear dynamics: application to piezoelectric-actuated nanopositioning stages. *IEEE-ASME Transactions on Mechatronics* 2019; 24(1): 326–337.
25. Wu L, Gao Y, Liu J, Li H. Event-triggered sliding mode control of stochastic systems via output feedback. *Automatica* 2017; 82(82): 79–92.
26. Li H, Shi P, Yao D, Wu L. Observer-based adaptive sliding mode control for nonlinear Markovian jump systems. *Automatica* 2016; 64(64): 133–142.
27. Yu S, Xie M, Wu H, Ma J, Wang R, Kang S. Design and control of a piezoactuated microfeed mechanism for cell injection. *The International Journal of Advanced Manufacturing Technology* 2019; 105(12): 4941–4952.
28. Yu S, Ma J,Wu H, Kang S. Robust precision motion control of piezoelectric actuators using fast nonsingular terminal sliding mode with time delay estimation. *Measurement Control* 2019; 52:11–19.
29. Fallaha C, Saad M, Kanaan HY, Alhaddad K. Sliding-mode robot control with exponential reaching law. *IEEE Transactions on Industrial Electronics* 2011; 58(2): 600–610.
30. Xie M, Li X, Wang Y, Liu Y, Sun D. Saturated PID control for the optical manipulation of biological cells. *IEEE Transactions on Control Systems Technology* 2018; 26(5): 1909–1916.
31. Xie M, Shakoor A, Li C, Sun D. Robust orientation control of multi-DOF cell based on uncertainty and disturbance estimation. *International Journal of Robust and Nonlinear Control* 2019; 29(14): 4859–4871.
32. Feng Y, Yu X, Man Z. Brief non-singular terminal sliding mode control of rigid manipulators. *Automatica* 2002; 38(12): 2159–2167.
33. Yu S, Yu X, Stonier RJ. Continuous finite-time control for robotic manipulators with terminal sliding modes. *Automatica* 2003; 41(11): 1957–1964.
34. Kang S, Wu H, Yang X, Li Y, Wang Y. Fractional-order robust model reference adaptive control of piezo-actuated active vibration isolation systems using output feedback and multiobjective optimization algorithm. *Journal of Vibrational and Control* 2020; 26(1–2): 19–35.
35. Lee J, Chang PH, Jin M. Adaptive integral sliding mode control with time-delay estimation for robot manipulators. *IEEE Transactions on Industrial Electronics* 2017; 64(8): 6796–6804.
36. Jin M, Lee J, Ahn KK. Continuous nonsingular terminal sliding-mode control of shape memory alloy actuators using time delay estimation. *IEEE-ASME Transactions on Mechatronics* 2015; 20(2): 899–909.

37. Ismail M, Ikhouane F, Rodellar J. The hysteresis Bouc–Wen model, a survey. *Archives of Computational Methods in Engineering* 2009; 16(2): 161–188.
38. Kilbas A, Srivastava H, Trujillo J. Fractional integrals and fractional derivatives. In: *Theory and Applications of Fractional Differential Equations*. Elsevier, *Berlin, Germany*, 2006, pp 69–133.
39. Levant A. Higher-order sliding modes, differentiation and output-feedback control. *International Journal of Control* 2003; 76:924–941.
40. Jia F, Cai X, Lou, Y, Li, Z. Interfacing technique and hardwarein- loop simulation of real-time co-simulation platform for wind energy conversion system. *IET Generation, Transmission & Distribution* 2017; 11(12): 3030–3038.
41. Lepanto P, Zolessi FR, Badano JL. Studying human genetic variation in zebrafish. In: *Cellular and Animal Models in Human Genomics Research*. Elsevier, *Berlin, Germany*, 2019, pp 89–117.
42. Nasevicius A, Ekker SC. Effective targeted gene 'knockdown' in zebrafish. *Nature Genetics* 2000; 26(2): 216–220.

6 Motion Tracking of Cell Puncture Mechanism Using Improved Sliding Mode Control with Time-Delay Estimation Technology

6.1 INTRODUCTION

Cell microinjection technology has introduced small doses of substances [1] (such as plasmids, RNA, antibodies, polypeptides, markers, inorganic ions, viruses, drugs, etc.) into active single cells through injection [2, 3]. This technology involves many precise cell operations, such as optical tweezers, electric field, magnetic field, dielectrophoresis, MEMS, infrasound field, and cell puncture [1, 4]. Cell microinjection is used widely by many biologists because of its high efficiency and convenience [5]. Cell puncture is the prerequisite and key step of cell microinjection. In cell puncture, an injection pipette with a diameter of less than 5 μm is used to penetrate the cell membrane and locates the injection pipette in specific cell tissue. The soft appearance of cells makes it prone to deformation, fragile vitality, small structure, and easy rupture of the cell membrane [6]. The biological structure characteristics of cells pose a serious challenge to cell puncture [7]. Cell puncture requires a clean environment and excellent operation skills [8]. It also requires submicron level operation accuracy and millisecond-level dynamic response time [3, 9]. Therefore, advanced mechatronics technology should be used to improve the intelligence of cell puncture, and a precise motion mechanism and the corresponding robust controller should be developed in dealing with complex working conditions, for instance, modeling uncertainty, unmodeled uncertainty, and external disturbance.

The cell puncture mechanism can be regarded as a complete ultra-high-precision biological robot system that requires a precise driving unit and a stable and reliable control system [10]. The key point to ensure precise motion is choosing an appropriate intelligent actuator [11]. Intelligent actuators have played an important role in many application fields with the development of high-precision and microminiaturized intelligent materials [12]. Piezoelectric ceramics (PEAs) have small structures, high resolution, fast response, and are noise- and pollution-free during operation [13], making them suitable for cell micromanipulation with high precision and environmental sensitivity [14]. However, the input voltage and output displacement of PEAs involve nonlinear mapping and their displacement range is small (usually less than 30 μm) because of the hysteresis and creep effect during their operation, which limits its movement considerably [15]. A displacement amplifying mechanism based on a compliant mechanism can connect with PEAs without gap and increase their displacement range. However, the hysteresis nonlinearity of PEAs will be aggravated. Thus, a robust motion controller should be developed to obtain high-precision motion.

Suppressing hysteresis nonlinearity is important to achieve high-precision motion. Control accuracy can be improved effectively by establishing an accurate dynamic model of the system with hysteresis. Hysteretic nonlinear models, including Preisach [16], Maxwell [17], Duhem [18], Prandtl-Ishlinskii [19], and Bouc–Wen [20] models, are used widely in literature. Time-consuming and laborious identification of dynamic parameters should be completed regardless of the hysteretic nonlinear model [21]. At the same time, the external disturbance, measurement noise, manufacturing error, and other factors in the actual operation of a system can cause a reduction in control accuracy. Considering the complexity of parameter identification and the existence of lumped disturbances, hysteretic nonlinearity is considered as an unknown term in the control strategy, thereby avoiding the parameter identification of the hysteretic nonlinearity. Robust controllers, such as proportional-integral-differential (PID) control, H∞ robust control, sliding mode control (SMC), and intelligent control, are designed based on a linear dynamic model [6]. In closed-loop control, sliding mode control, which is an efficient and stable robust controller, has remarkable advantages in dealing with the uncertainties and external disturbances of a system [7, 22].

The traditional SMC usually uses a linear hyperplane, namely, a PD-type sliding surface. However, this widely used PD-type sliding surface has slow response speed. The integral sliding surface can reduce the steady-state error of the controller [23]. Meanwhile, an improved PID-type sliding surface based on a PD-type sliding surface provides fast response and precise control [24]. Therefore, SMC with a PID-type sliding surface has fast transient response and has been used widely by many scholars in the control of nonlinear systems. The sign functions in traditional SMC are used typically to cause the state trajectories of the system to intersect and remain on the manifold. A sign function is discontinuous, causing chattering to the system. Buffeting causes high-frequency switching of circuit signals, hardware heating, and short service life. Mechanical shock and self-excited vibration occur in mechanical systems [25]. However, the control law can be changed from discontinuous to continuous output by introducing a fast reaching law, thereby eliminating chattering [26]. FPIDSMC is composed of SMC that combines fast reaching law and PID-type sliding surface. It tends to be conservative when choosing controller parameters because the upper limit of the unknown system boundary is not known, thereby reducing control accuracy. Therefore, unknown terms should be estimated in real time to improve the control accuracy of FPIDSMC.

Time-delay estimation (TDE) is used considerably in model-free control because it can estimate the unknown variables of a system easily and efficiently [27]. TDE uses the time-delay information of state variable and control law to estimate unknown terms. It assumes that the change in unknown quantities is small under a sufficiently small time period (also known as time-delay parameter L) [28]. TDE technology is used to realize online estimation and compensation of unknown quantities and reduce the gain of FPIDSMC. Meanwhile, FPIDSMC and TDE technology are combined to form an FPID-TDE controller that makes their advantages become perfectly complementary, allowing the construction of a controller with strong robustness and high control accuracy.

Two preconditions should be accomplished to obtain ideal estimation accuracy in the practical application of TDE technology, which are described as follows: 1) delay parameter L should be sufficiently small and 2) the feedback of velocity and acceleration should be sufficiently accurate [29]. Unfortunately, the two preconditions are difficult to meet perfectly because of limited computer operation speed, sensor sampling frequency, and communication speed and because the delay parameter L cannot be infinitely small. In an actual operation, only a displacement sensor is arranged, and the speed or acceleration signal is obtained through differential operation of the displacement signal, leading to error accumulation. Therefore, it needs not only to rely on TDE technology to realize the estimation of the unknown term but also to reduce the estimation error as much as possible. This problem can be solved skillfully by reducing the proportion of unknown items in a system. A system dynamic equation based on the Bouc–Wen model includes linear and hysteretic nonlinear parts. The linear part is identified through engineering and the hysteretic nonlinear part is classified as the unknown term, which can reduce effectively the proportion of unknown terms in the dynamic model. Therefore, this chapter presents an FPID-TDE controller based on the Bouc–Wen model that can estimate and suppress the unknown term in the limited error range.

The main contributions of this chapter are threefold. First, the nonlinear part of hysteresis is classified as the unknown term, thereby reducing the complexity of the model and making the model-based FPID-TDE controller simple and convenient for engineering applications. Second, the FPID-TDE controller is designed based on the Bouc–Wen model, and the stability analysis of the controller is verified based on the Lyapunov theory. The proposed SMC with PID-type sliding surface and fast reaching law has small gain, continuous output, high control accuracy, and strong robustness. The estimation and compensation of unknown terms realized through TDE technology are simple and efficient. Finally, the performance of the FPID-TDE controller is verified through comparisons with multiple controllers and a puncture test of zebrafish embryo is conducted. The results show the FPID-TDE controller is effective and reliable.

The rest of this chapter is arranged as follows. Section 6.2 presents the controller design, parameter debugging of the robust precise differentiator, and stability analysis of the controller. Section 6.3 conducts computer simulation experiments for the proposed controller. Section 6.4 shows the semi-physical simulation experiments and then performs the cell micropuncture experiment of a zebrafish embryo. Section 6.5 concludes this chapter.

6.2   ROBUST CONTROLLER DESIGN

This section focuses on the design of a robust controller and stability analysis based on the simplified system dynamics formulated in Chapter 5.

6.2.1   CONTROLLER DESIGN

The control system aims to realize that the output displacement x of the cell puncture mechanism can track the expected displacement x_d accurately. The displacement error equation and its derivative (i.e., velocity error equation) are defined as follows: $e = x_d - x$ and $\dot{e} = \dot{x}_d - \dot{x}$. The dynamic equation of target error is given as

$$\ddot{e} + K_P e + K_I \dot{e} = 0 \tag{6.1}$$

where K_I and K_P are the constant positive parameters to be adjusted. The above goal is equivalent to tracking the following PID-type sliding surface, i.e. $s = 0$:

$$s = \dot{e} + K_P e + K_I \int e\, d\tau \tag{6.2}$$

with the proportioning coefficient $K_P > 0$ and integral coefficient $K_I > 0$.

The SMC controller with fast reaching law and PID-type sliding surface is designed as follows:

$$u = \frac{1}{kd}(u_1 + u_2) \tag{6.3}$$

where $u_1 = m\left(u_0 + \beta \mathrm{sig}(s)^\rho\right) + b\dot{x} + kx$, $u_2 = -m\Delta\hat{P}$, $u_0 = \ddot{x}_d + K_P\dot{e} + K_I e$, gain $\beta > 0$, and $0 < \rho < 1$.

The expression is simplified as $\mathrm{sig}(s)^\rho = |x|^\rho \mathrm{sign}(s)$, where $\Delta\hat{P}(t)$ is the estimate of $\Delta P(t)$. In this chapter, we use TDE technology to obtain $\Delta\hat{P}(t)$ online. $\Delta\hat{P}(t)$ is expressed as follows:

$$\Delta\hat{P}(t) = \Delta P(t-L) = \ddot{x}(t-L)$$
$$-\frac{kd}{m}u(t-L) + \frac{b}{m}\dot{x}(t-L) + \frac{k}{m}x(t-L) \tag{6.4}$$

where L is the delay parameter and $\Delta P(t - L)$ is the value of ΔP at time $t - L$. The smaller the value of delay parameter L, the closer $\Delta P(t)$ and $\Delta P(t-L)$. The minimum value that L can set is the sampling time of the computer system. The error of TDE is defined as $\Delta\tilde{P}(t) = \Delta\hat{P}(t) - \Delta P(t)$.

Remark 1:

The implementation of TDE depends on the controller output and state variables at the previous time to estimate the unknown term at the current time. TDE technology is simple and easy to understand. However, TDE technology can only guarantee the gradual convergence of system tracking error and cannot ensure the control quality and external disturbance robustness of the system.

The expression of FPIDSMC control law based on TDE technology can be obtained by substituting Eq. (6.4) into (6.3) as follows:

$$
u_{\text{FPID TDE}} = \underbrace{\frac{1}{kd}\left\{ m\left[\ddot{x}_d + K_P\dot{e} + K_I e + \beta \operatorname{sig}(s)^\rho \right] + b\dot{x} + kx \right\}}_{\text{FPIDSMC}} \tag{6.5}
$$
$$
\underbrace{-\frac{m}{kd}\left[\ddot{x}(t-L) - \frac{kd}{m}u(t-L) + \frac{b}{m}\dot{x}(t-L) + \frac{k}{m}x(t-L) \right]}_{u_{\text{TDE}}}
$$

The controller combining FPIDSMC and TDE technology is abbreviated as FPID-TDE.

Remark 2:

Gain β of FPID-TDE can be reduced immensely because TDE technology is used to estimate and compensate for the unknown term, thereby ensuring the stability of the control law output.

A closed-loop displacement error equation can be obtained by substituting control law (6.1) into the system dynamics Eq. (6.5):

$$
\ddot{e} + K_P\dot{e} + K_I e + \beta \operatorname{sig}(s)^\rho - \Delta\tilde{P}(t) = 0 \tag{6.6}
$$

Remark 3:

As shown in the closed-loop displacement error of Eq. (6.6), displacement error e is independent of the expected displacement and disturbance. In particular, the disturbance acting on the system cannot affect the convergence rate of displacement error e regardless of its type. It embodies the strong robustness of FPID-TDE.

6.2.2 ESTIMATION OF FULL STATE FEEDBACK

As shown in the control law of Eq. (6.5), the entire state trajectory (expected displacement, expected velocity, and expected acceleration) and the entire state feedback (actual displacement, actual velocity, and actual acceleration) are needed in the realization. The expected velocity and acceleration can be obtained through differential calculation of the expected displacement in advance. Unfortunately, only displacement sensors are used in practical applications. Thus, the actual velocity and acceleration must be estimated by the actual displacement. The noise signal will be amplified directly because the actual displacement contains noise signal in terms of differential operation or discrete difference operation. In this chapter, a robust exact differentiator (RED) [30] is selected to realize the accurate estimation of full state feedback. RED is expressed as follows:

$$
\begin{cases}
\dot{z}_0 = v_0 = -\lambda_1 |z_0 - x|^{2/3} \, sign(z_0 - x) + z_1 \\
\dot{z}_1 = v_1 = -\lambda_2 |z_1 - v_0|^{1/2} \, sign(z_1 - v_0) + z_2 \\
\dot{z}_2 = -\lambda_3 \, sign(z_2 - v_1)
\end{cases}
\tag{6.7}
$$

where $\lambda_1 = 3\lambda^{1/3}$, $\lambda_2 = 1.5\lambda^{1/2}$, $\lambda_3 = 1.2\lambda$, $\lambda \geq |x|$, $z_0 = \hat{x}$, $z_1 = \dot{\hat{x}}$, $z_2 = \ddot{\hat{x}}$ are the estimation of the expected displacement, expected velocity, and expected acceleration, respectively.

Next, the performance of RED will be revealed through simulation experiments. RED is used to estimate the three given signals, and the expected values are expressed as follows:

Expected displacement $x = 10 \sin(4\pi t e^{-0.2t})$,

Expected velocity $\dot{x} = -(8\pi t - 40\pi)e^{-0.2t} \cos(4\pi t e^{-0.2t})$ and expected accelera-

tion $\ddot{x} = -0.4e^{-0.4t} \left[\begin{array}{l} (16\pi^2 t^2 - 160\pi^2 t + 400\pi^2)\sin(4\pi t e^{-0.2t}) \\ + (40\pi - 4\pi t)e^{0.2t} \cos(4\pi t e^{-0.2t}) \end{array} \right]$.

White noise is added to the expected values during simulation to reflect comprehensively the performance of RED. The $\lambda = 0.01$ is selected to estimate the expected values and obtain the estimated values. As shown in Figure 6.1 (a) and (c), RED can estimate the displacement and velocity signals accurately and reduce the effects of white noise signals effectively on estimation accuracy. Figure 6.1 (e), the estimated value of the acceleration signal is barely close to the expected value, and its estimation accuracy is reduced considerably. As shown in Figure 6.1 (b), (d), and (f), the accuracy is reduced gradually because error accumulation is inevitable in the estimation of RED. The accuracy of the acceleration estimation value is less than that of the velocity estimation value because the estimation of the acceleration signal depends on the velocity signal. Therefore, the influence of full state feedback error on TDE technology should be considered.

FPIDSMC and TDE technology are synthesized and RED is used to estimate the full state feedback. The control flowchart of FPID-TDE is shown in Figure 6.2.

6.2.3 STABILITY ANALYSIS

Proof: The Lyapunov function is expressed as follows:

$$
V = \frac{1}{2} s^2
\tag{6.8}
$$

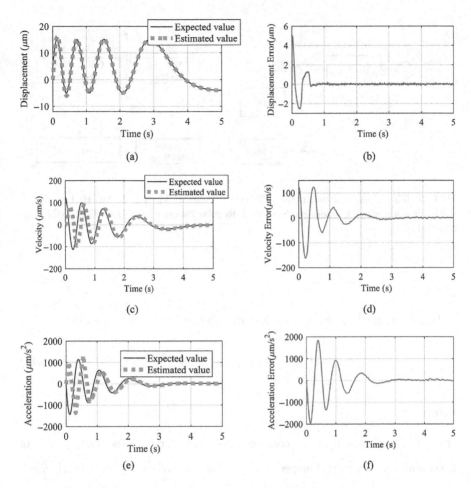

FIGURE 6.1 RED is used to realize full state feedback estimation and its error curve: (a) estimation of displacement signal, (b) estimation error ($e_0 = \hat{x} - x$) of expected displacement ($z_0 = \hat{x}$), (c) estimation of velocity signal, (d) estimation error ($e_1 = \hat{\dot{x}} - \dot{x}$) of expected velocity ($z_1 = \dot{\hat{x}}$), (e) estimation of acceleration signal, and (f) estimation error ($e_2 = \hat{\ddot{x}} - \ddot{x}$) of expected acceleration ($z_2 = \ddot{\hat{x}}$). (Source: M. Xie, S. Yu, H. Lin et al./IEEE Transactions on Circuits and Systems I: Regular Papers 67 (9) 3199–3210, 2022, with permission.)

For the error of TDE $\Delta\tilde{P}$, a positive number of φ is found, which makes $\Delta\tilde{P}$ bounded [31].

$$\left|\Delta\tilde{P}\right| \leq \phi \tag{6.9}$$

Eq. (6.6) is re-expressed as $\dot{s} = \Delta\tilde{P} - \beta\,\mathrm{sig}\left(s\right)^{\rho}$.

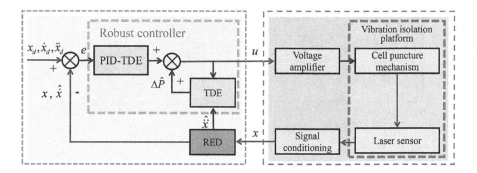

FIGURE 6.2 Control flowchart of FPID-TDE. (Source: M. Xie, S. Yu, H. Lin et al. / IEEE Transactions on Circuits and Systems I: Regular Papers 67 (9) 3199–3210, 2022, with permission.)

Differentiating Lyapunov function (6.8) with respect to time [32] yields:

$$\dot{V} = s\dot{s} = s\left(\Delta\tilde{P} - \beta\,\mathrm{sig}(s)^{\rho}\right) \le -\beta|s|\left(|s|^{\rho} - \phi\big/\beta\right) \tag{6.10}$$

Remark 4:

The FPID-TDE controller has fast convergence and high accuracy. The state trajectory of the system will converge to the specified region, which can be expressed as $|s| \le \left(\phi\big/\beta\right)^{\gamma\big/\rho}$.

6.2.4 Parameter Tuning of the Controller

In the FPID-TDE controller, the adjustable parameters are β, ρ, K_P, K_I, and L.

Remark 5:

Eq. (6.5) shows that β can affect the convergence rate of the system on the sliding surface and that a small β can achieve small gain and short convergence time.

Remark 6:

Eq. (6.5) shows that the control law can achieve continuous output without chattering and have strong robustness because the exponent ρ is less than 1 and more than 0. Different exponents ρ can enable control laws work in different modes. For example, the controller is linear when $\rho \to 1$, whereas the controller is discontinuous when $\rho \to 0$.

Remark 7:

Eq. (6.2) shows that K_p reflects displacement error e of the control system proportionally and accelerates its convergence speed. The steady-state error of the system will disappear depending on the integral term K_I. The design of K_p and K_I makes the sliding mode stable at $s = 0$, that is, e and $\dot{e}$ converge to 0 when s converges to 0, which denotes K_p and K_I should be positive values. Otherwise, L can be set as an integral multiple of the sampling period, and the smaller the value L is, the higher the control accuracy is.

6.3   COMPUTER SIMULATION EXPERIMENT

The proposed FPID-TDE controller is a model with a PID-type sliding surface and fast reaching law. TDE technology is used to estimate the unknown quantity. It has a generalized structure and can be evolved into other types of controllers.

6.3.1   CONTROLLERS FOR COMPARISON

First, we derive the three existing controllers.

Controller a: Traditional PD-type sliding surface controller
   The PID-type sliding surface will become PD-type sliding surface when $K_I = 0$,
   that is, $s = \dot{e} + K_p e$. The proposed controller will evolve as follows:

$$u_{\text{FPD-TDE}} = \underbrace{\frac{1}{kd}\left\{m\left[\ddot{x}_d + K_p\dot{e} + \beta\,\text{sig}(s)^\rho\right] + b\dot{x} + kx\right\}}_{\text{FPD-SMC}}$$

$$\underbrace{-\frac{m}{kd}\left[\ddot{x}(t-L) - \frac{kd}{m}u(t-L) + \frac{b}{m}\dot{x}(t-L) + \frac{k}{m}x(t-L)\right]}_{u_{\text{TDE}}} \quad (6.11)$$

The above controller is abbreviated as an FPD-TDE controller for simplicity.

Controller b (Jin's controller)

Referring to the results in [33], the system dynamics model of the cell puncture mechanism is transformed into the following:

$$\bar{m}\ddot{x} + N\left(x,\dot{x},\ddot{x},k,d,h,\tau_d\right) = u \quad (6.12)$$

The unknown item is $N\left(x,\dot{x},\ddot{x},k,d,h,\tau_d\right) = \left(\frac{m}{kd} - \bar{m}\right)\ddot{x} + \frac{1}{k}\dot{x} + \frac{1}{d}x$
$+ \frac{h}{d} - \frac{\tau_d}{kd}$. $N\left(x,\dot{x},\ddot{x},k,d,h,\tau_d\right)$ is abbreviated as $N(t)$ for simplicity. Eq. (6.12) is simplified as follows:

$$\bar{m}\ddot{x} + N(t) = u \quad (6.13)$$

The model-free controller can be determined by following the design idea of the FPID-TDE controller using a dynamic model (6.12).
TDE technology is used to estimate the unknown term $N(t)$:

$$\hat{N}(t) = N(t-L) = u(t-L) - \bar{m}\ddot{x}(t-L) \tag{6.14}$$

The controller is designed as follows:

$$u_{\text{Jin}} = \underbrace{\bar{m}\left(\ddot{x}_d + K_p\dot{e} + K_I e + \beta \operatorname{sig}(s)^\rho\right)}_{\text{FPIDSMC}} + \underbrace{u(t-L) - \bar{m}\ddot{x}(t-L)}_{u_{\text{TDE}}} \tag{6.15}$$

This controller was proposed by Maolin Jin in 2009. The controller is abbreviated as Jin's controller [33].

Remark 8:

The advantages of model-free control are that it can omit the construction of a dynamic model and parameter identification and can be implemented easily in engineering. However, the value of the unknown term $N(t)$ is greater than $\Delta P(t)$ because the "information" contained in the dynamic model (6.13) is less than that in the dynamic model (5.5).

Controller c (Hsia's controller)
In Jin's controller, the fast reaching law will be eliminated, and Jin's controller will become Hsia's position controller when $\beta = 0$, which can be expressed as

$$u_{\text{Hsia}} = \underbrace{\bar{m}\left(\ddot{x}_d + K_p\dot{e} + K_I e\right)}_{\text{PIDSMC}} + \underbrace{u(t-L) - \bar{m}\ddot{x}(t-L)}_{u_{\text{TDE}}} \tag{6.16}$$

The above controller is abbreviated as Hsia's controller [34].
The delay estimation errors of controllers b and c are defined as follows:

$$\tilde{N}(t) = \hat{N}(t) - N(t-L) \tag{6.17}$$

Remark 9:

Jin's and Hsia's controllers have concise structure, simple design, and small calculation. In comparison, the FPID-TDE controller has the largest amount of computation. The step size increases when the FPIF-TDE controller needs to run in a single-chip system.

6.3.2 COMPUTER SIMULATION EXPERIMENTS

The comparative experiment of the four controllers is conducted on a computer, with a CPU speed of 3.6 GHz and Windows 7 operating system. The simulation software is MATLAB Simulink 2016b and the step size is 0.1 ms. The expected displacement used in the displacement tracking experiment is a typical sine curve with an amplitude of 100 μm and a frequency of 0.5 Hz (as shown in Figure 6.3). The parameters of the controller are listed in Table 6.1. The values of relevant parameters are the same.

Root mean square error (*RMSE*) and maximum error (*ME*) [35] are selected as the inspection indices defined as follows:

$$RMSE = \sqrt{\sum_{I=1}^{N} \left(x_{di} - x_i \right)^2 \Big/ N} \qquad (6.18)$$

$$ME = \max \left(\left| x_{di} - x_i \right| \right) \qquad (6.19)$$

where N is the total number of samples, and x_{di} and x_i are the expected and actual values of the ith sampling signal, respectively.

Four controllers can achieve accurate tracking of the sinusoidal signal. Table 6.2 and Figure 6.4 show that the performance of FPID-TDE and Jin's controllers are consistent in the computer simulation. The computer simulation has no external

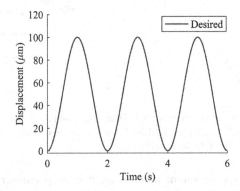

FIGURE 6.3   Sine displacement curve with constant amplitude and frequency. (Source: M. Xie, S. Yu, H. Lin et al./IEEE Transactions on Circuits and Systems I: Regular Papers 67 (9) 3199–3210, 2022, with permission.)

TABLE 6.1
Controller Parameter Design

Controllers	Parameter values
FPID-TDE	$\beta = 13, \rho = 0.6, K_P = 200, K_I = 10^5$;
FPD-TDE	$\beta = 13, \rho = 0.6, K_P = 200, K_I = 0$;
Jin's controller	$\beta = 13, \rho = 0.6, K_P = 200, K_I = 10^5$;
Hsia's controller	$\beta = 0, \rho = 0, K_P = 200, K_I = 10^5$;

Source: M. Xie, S. Yu, H. Lin et al./IEEE Transactions on Circuits and Systems I: Regular Papers 67 (9) 3199–3210, 2022, with permission.

TABLE 6.2

Inspection Index List of Four Controllers

Performance Indicators	Parameter values			
	FPID-TDE	FPD-TDE	Jin's controller	Hsia's controller
RMSE (μm)	2.2×10^{-5}	4.6×10^{-5}	2.2×10^{-5}	7.7×10^{-5}
ME (μm)	0.031	0.101	0.031	1.062

Source: M. Xie, S. Yu, H. Lin et al./IEEE Transactions on Circuits and Systems I: Regular Papers 67 (9) 3199–3210, 2022, with permission.

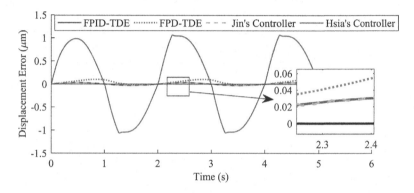

FIGURE 6.4   Displacement error curve when tracking a sinusoidal displacement curve of 0.5 Hz. (Source: M. Xie, S. Yu, H. Lin et al./IEEE Transactions on Circuits and Systems I: Regular Papers 67 (9) 3199–3210, 2022, with permission.)

interference and sensor measurement noise and thus the parameter setting of the controller is the same. The only difference is that the former is model-based control, and the latter is model-free control. However, this condition is not the basis of their performance.

Why does the model-free control achieve the same control effect as that of a model-based controller? The four curves in Figure 6.5 should be observed and compared to explain this condition clearly.

Remark 10:

In the FPID-TDE controller, TDE technology realizes accurate estimation of the unknown term when the estimation error of TDE is 3 mV. In Jin's controller, the estimation error of TDE is 10 mV. The output of the control law is approximately 60 V to drive the cell puncture mechanism to 100 μm. Therefore, the estimation error of 10 mV is negligible as compared to the control voltage of 60 V because the two controllers can estimate and compensate accurately for the unknown term. Under the ideal environment of computer simulation, the two controllers can achieve the same control effect.

A comparison of the displacement curves in Figure 6.4 shows that the control effect of FPD-TDE is worse than that of FPID-TDE because of the lack of integration link.

Remark 11:

The integral term has a good inhibition effect on the accumulation of slowly changing errors.

Figure 6.4 further shows that compared with Jin's controller, Hsia's controller has the lowest accuracy because of its lack of fast reaching law.

Figure 6.5 shows the error curve of TDE in four controllers used to evaluate the estimation accuracy of TDE. The accuracy of TDE is obviously improved with the decrease in unknown terms in the dynamic model. The TDE error is negligible relative to the output of control law.

Remark 12:

TDE technology can estimate the unknown term accurately based on the small step size (0.1 ms) and the performance of robust precise differentiator. This condition shows indirectly that the designed robust precise differentiator is reliable and effective.

Apart from the comparison of FPID-TDE and three controllers, the output of their control law includes two parts, namely, sliding mode and TDE parts. We will explore the extent of the two parts.

Figure 6.6 shows the control law output of four controllers, including the total control law output curve and control law output (u_{TDE}) caused by TDE. Figure 6.6 (a) and (b) show that the TDE has a limited action intensity compared with the total control law output and that the sliding mode part plays a huge role. Figure 6.6 (c) and (d) indicate that TDE technology (u_{TDE}) plays an absolute role.

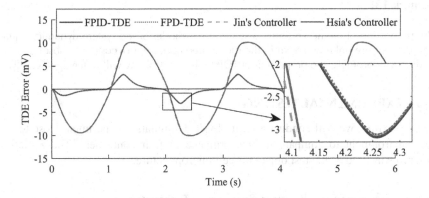

FIGURE 6.5 Error curve of unknown term estimation using TDE Technology. (Source: M. Xie, S. Yu, H. Lin et al./IEEE Transactions on Circuits and Systems I: Regular Papers 67 (9) 3199–3210, 2022, with permission.)

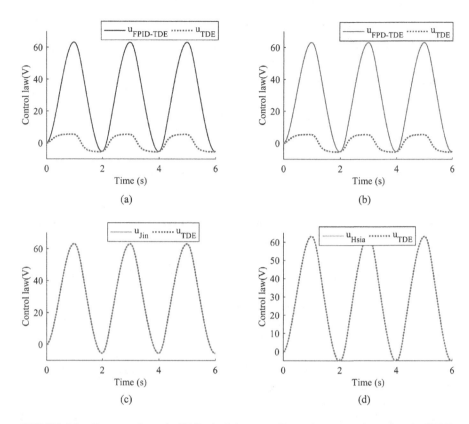

FIGURE 6.6   Output voltage (solid line) of the controller and output voltage (u_{TDE}) of TDE. (a) FPID-TDE, (b) FPD-TDE, (c) Jin's controller, (d) Hsia's controller. (Source: M. Xie, S. Yu, H. Lin et al./IEEE Transactions on Circuits and Systems I: Regular Papers 67 (9) 3199–3210, 2022, with permission.)

Remark 13:

The proportion of unknown terms is small in the model-based system dynamics equation design. The unknown term in the model-free controller occupies an absolute proportion. Accordingly, the proportion of TDE in the control law will differ significantly.

6.4   EXPERIMENTAL TESTING

In this section, we will conduct a semi-physical simulation experiment close to the real environment to compare the performance of four controllers. The zebrafish embryo was used as the test object for cell micropuncture.

6.4.1   SETUP OF SEMI-PHYSICAL SIMULATION EXPERIMENT

Figures 6.7 and 6.8 show the high-performance host computer and target computer prototype environment used with xPC Target as the core technology. The host computer runs MATLAB®/Simulink® 2016b and the target computer runs in DOS XPC mode

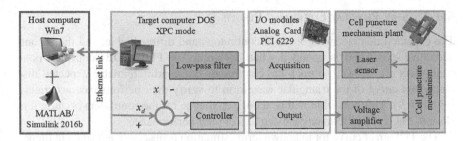

FIGURE 6.7 System block diagram of the cell puncture mechanism based on the XPC target mode. (Source: M. Xie, S. Yu, H. Lin et al./IEEE Transactions on Circuits and Systems I: Regular Papers 67 (9) 3199–3210, 2022, with permission.)

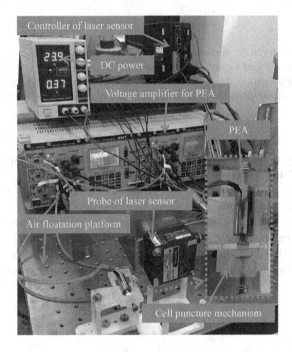

FIGURE 6.8 Semi-physical simulation experimental device of the cell puncture mechanism. (Source: M. Xie, S. Yu, H. Lin et al./IEEE Transactions on Circuits and Systems I: Regular Papers 67 (9) 3199–3210, 2022, with permission.)

[35]. XPC Target can connect the Simulink model with the physical system and use the real-time workshop to generate codes automatically and run on the target computer.

The main core components of the experimental device include bridge-type displacement amplification mechanism (the material is a mixture of photosensitive resin and polypropylene fabricated through 3D printing), PEA (model Pst120/7/20vs12, from Harpin Core Tomorrow Technology Co., Ltd), laser displacement sensor (model LK-H020, from KEYENCE Corporation with a measurement accuracy is 0.02 μm), and analog signal card (model PCI-6229, from National Instruments Corporation). The step size of the experimental system is set to 1 ms.

6.4.2 SEMI-PHYSICAL EXPERIMENT RESULTS

The expected displacement curve is continuous and differentiable, and the second derivative exists and is bounded when designing the FPID-TDE controller. Four controllers are allowed to track the continuous but nondifferentiable expected displacement curve of the triangular waveform to verify their performance accurately (Figure 6.9). This condition will be a huge challenge for the controllers because it is close to the real environment and can reflect their performance.

The TDE error curve for unknown term estimation is observed, as shown in Figure 6.10. The TDE errors of the four controllers are larger compared with the computer simulation environment. These errors cannot be ignored relative to the displacement error.

Remark 14:

In practice, the estimation accuracy of TDE technology will be reduced because of the increase in step size and the influence of noise. Therefore, setting a minimum step size, applying robust precision differentiator with excellent performance, and adding noise filtering device in the processing of feedback signals are effective means of improving TDE technology.

The displacement error curves of the four controllers in the semi-physical experiment are shown in Figure 6.11. All four controllers can track the expected displacement curve of the triangle waveform. Among them, FPID-TDE control has the highest tracking accuracy, and Jin and Hsia's controllers have the lowest tracking accuracy. The displacement error curves of FPD-TDE and FPID-TDE do not fluctuate evidently with the expected displacement curves of triangle wave. On the contrary, the tracking errors of Jin's and Hsia's controllers increase at the turning point of the expected displacement curve of the triangle waveform. The performance index of Jin's controller is worse than that of the proposed controller from the indexes of

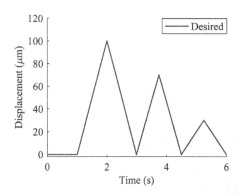

FIGURE 6.9 Expected displacement curve of the triangular waveform. (Source: M. Xie, S. Yu, H. Lin et al./IEEE Transactions on Circuits and Systems I: Regular Papers 67 (9) 3199–3210, 2022, with permission.)

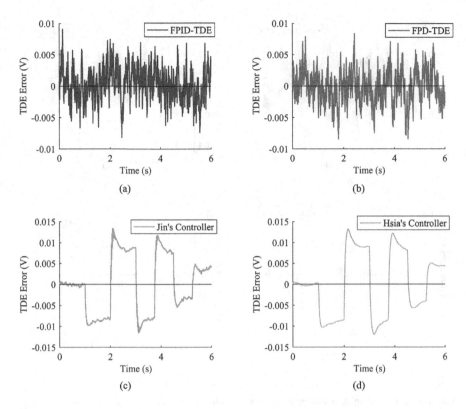

FIGURE 6.10   TDE error curve of unknown term estimation is realized by TDE technology: (a) FPID-TDE, (b) FPD-TDE, (c) Jin's controller, (d) Hsia's controller. (Source: M. Xie, S. Yu, H. Lin et al./IEEE Transactions on Circuits and Systems I: Regular Papers 67 (9) 3199–3210, 2022, with permission.)

RMSE and *ME*. The variation trend of the error curve estimated by the unknown term in Figure 6.10 (c) and (d) is similar to that of the displacement error in Figure 6.11 (c) and (d).

Remark 15:

The outputs of the control laws of Jin's and Hsia's controllers in the computer simulation experiment are realized by TDE technology. In the semi-physical environment, the estimation accuracy of TDE technology decreases more than that of the FPD-TDE and FPID-TDE controllers and affects the amplification of displacement error.

The semi-physical environment reduces the estimation accuracy of TDE technology compared with computer simulation. The model-based controller can reduce the estimation error of TDE technology by reducing the proportion of unknown terms in the system dynamics model to achieve good estimation and compensation of unknown terms using TDE technology. FPID-TDE has high control accuracy, and its control law has continuous output, smooth, without chattering, and strong robustness.

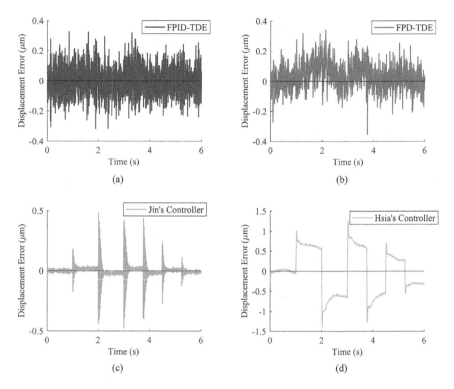

FIGURE 6.11 Displacement error curves of four controllers in the semi-physical experiment. (a) FPID-TDE, (b) FPD-TDE, (c) Jin's controller, (d) Hsia's controller. (Source: M. Xie, S. Yu, H. Lin et al./IEEE Transactions on Circuits and Systems I: Regular Papers 67 (9) 3199–3210, 2022, with permission.)

6.4.3 CELL PUNCTURE EXPERIMENT

Given the great technical demand for zebrafish embryo injection in the medical field, this chapter uses a zebrafish embryo as an example in conducting the cell puncture experiment. Zebrafish is the preferred model organism because it has 87% homology with human genes. The experimental results of zebrafish are mostly suitable for the human body. Cell labeling, tissue transfer, mutation, haploid breeding, transgenic, and gene activity inhibition technologies of zebrafish have increasingly developed [36]. Thousands of zebrafish embryo mutants are available, which are excellent resources to study the molecular mechanism of embryonic development, and some of them can be used as human disease models. Zebrafish embryos are transparent, and the location of the injection pipette in the cell can be clearly observed.

The experimental setup of a zebrafish embryo puncture is shown in Figure 6.12. The cell puncture mechanism is installed in an inverted biological microscope (BA1000, Chongqing Optical Instrument Factory). An electron microscope (21 million pixels, Shenzhen ZongyuanWeiye Technology Co., Ltd.) is installed in the upper part to enable the observation of the puncturing process. The micropuncture process of the zebrafish embryo is shown in Figure 6.13. The results of 20 experiments show that the time to complete a puncture is stable within 0.6 s.

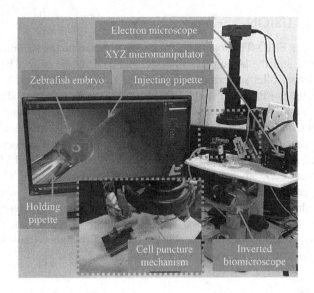

FIGURE 6.12 Puncture experiment of zebrafish embryo using the cell puncture mechanism. (Source: M. Xie, S. Yu, H. Lin et al./IEEE Transactions on Circuits and Systems I: Regular Papers 67 (9) 3199–3210, 2022, with permission.)

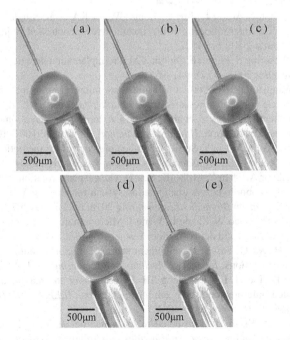

FIGURE 6.13 Micropuncture of zebrafish embryos. (a) Injection pipette reaches the matched cell membrane of zebrafish, (b), (c) injection pipette gradually compresses and deforms the cell membrane, (d) injection pipette enters the cell membrane, (e) injection pipette exits the zebrafish embryo. (Source: M. Xie, S. Yu, H. Lin et al./IEEE Transactions on Circuits and Systems I: Regular Papers 67 (9) 3199–3210, 2022, with permission.)

6.5 CONCLUSION

In this chapter, a FPID-TDE controller is designed based on the simplified dynamic model of CPM and the advantages of SMC and TDE technology. The stability of the proposed controller is verified based on the Lyapunov theory. The FPID-TDE controller includes FPIDSMC and TDE. FPIDSMC has a PID-type sliding surface and fast reaching law that can achieve a fast response, few steady-state errors, continuous output, and without chattering. The TDE technology realizes accurate estimation and compensation of unknown terms. Thus, the controller does not require prior knowledge of an unknown disturbance boundary. FPIDSMC and TDE combine and complement each other, with FPIDSMC reducing the burden of TDE and TDE reducing the gain of FPIDSMC. Computer simulation and semi-physical simulation experiments show the proposed FPID-TDE controller has high efficiency, high control accuracy, and strong robustness. The cell puncture mechanism and FPID-TDE controller are applied to the micropuncture of a zebrafish embryo and was observed to be capable of completing a cell puncture within 0.6 s.

REFERENCES

1. Permana S, Grant E, Walker GM, Yoder JA. A review of automated microinjection systems for single cells in the embryogenesis stage. *IEEE/ASME Transactions on Mechatronics* 2016; 21(5): 2391–2404.
2. Chow YT, Chen S, Wang R, Liu C, Kong CW, Li RA et al. Single cell transfection through precise microinjection with quantitatively controlled injection volumes. *Scientific Reports* 2016; 6: 24127.
3. Braude P, Pickering S, Flinter F, Ogilvie CM. Preimplantation genetic diagnosis. *Nature Reviews Genetics* 2002; 3(12): 941–955.
4. Zhao H, Wang S. Different effect of l-glutamate microinjection into medial or lateral habenular nucleu on blood pressure. *Acta Physiologica Sinica* 1995; 47(3): 292.
5. Van Steirteghem AC, Nagy Z, et al. High fertilization and implantation rates after intracytoplasmic sperm injection. *Human Reproduction.* 1993; 8(7): 1061–1066.
6. Xie M, Li X, Wang Y, Liu Y, Sun D. Saturated pid control for the optical manipulation of biological cells. *IEEE Transactions on Control Systems Technology* 2017; (26): 1–8.
7. Xie M, Shakoor A, Shen Y, Mills JK, Sun D. Out-of-plane rotation control of biological cells with a robot-tweezers manipulation system for orientation-based cell surgery. *IEEE Transactions on Biomedical Engineering* 2018; 66(1): 199–207.
8. Li J, Berta EFDÁ, Gao W, Zhang L, Wang J. Micro/nanorobots for biomedicine: delivery, surgery, sensing, and detoxification. *Science Robotics* 2017; 2(4): eaam6431.
9. Li Y, Tang H, Xu Q, Yun Y. Development status of micromanipulator technology for biomedical applications. *Journal of Mechanical Engineering* 2011; 47(23): 1–13.
10. Xie Y, Sun D, Tse H, Liu C, Cheng SH. Force sensing and manipulation strategy in robot-assisted microinjection on zebrafish embryos. *IEEE/ASME Transactions on Mechatronics* 2011; 16(6): 1002–1010.
11. Tang Y, Gao H, Kurths J. Distributed robust synchronization of dynamical networks with stochastic coupling. *IEEE Transactions on Circuits & Systems I Regular Papers* 2017; 61(5): 1508–1519.
12. Yu S, Ma J, Wu H, Kang S. Robust precision motion control of piezoelectric actuators using fast nonsingular terminal sliding mode with time delay estimation. *Measurement and Control* 2018; 52(1–2): 11–19.

13. Ronkanen P, Kallio P, Vilkko M and Koivo HN. Displacement control of piezoelectric actuators using current and voltage. *IEEE/ASME Transactions on Mechatronics* 2011; 16(1): 160–166.

14. Sun D, Sheng W, Hata S, Sakurai J, Shimokohbe A. Theoretical and experimental investigation of traveling wave propagation on a several-millimeter-long cylindrical pipe driven by piezoelectric ceramic tubes. *IEEE Transactions on Ultrasonics Ferroelectrics and Frequency Control* 2010; 57(7): 1600–1611.

15. Liu P, Yan P, Zhang Z, Leng T. Modeling and control of a novel x-y parallel piezoelectric-actuator driven nanopositioner. *ISA Transactions* 2015; 56: 145–154.

16. Wei YD, Tao HF. Study the Preisach model of hysteresis in piezoelectric actuator. *Piezoelectrics & Acoustooptics* 2004; 26: 364–367.

17. Liu Y, Liu H, Wu H, Zou D. Modelling and compensation of hysteresis in piezoelectric actuators based on Maxwell approach. *Electronics Letters* 2016; 52(3): 188–190.

18. Oh JH, Bernstein DS. Semilinear Duhem model for rate-independent and rate-dependent hysteresis. *IEEE Transactions on Automatic Control* 2005; 50(5): 631–645.

19. Kuhnen K. Modeling, identification and compensation of complex hysteretic nonlinearities: a modified prandtl-ishlinskii approach. *European Journal of Control* 2003; 9(4): 407–418.

20. Ismail M, Rodellar J, Ikhouane F. Performance of structure–equipment systems with a novel roll-n-cage isolation bearing. *Computers & Structures* 2009; 87(23–24): 1631–1646.

21. Wen Z, Ding Y, Liu P, Ding H. An efficient identification method for dynamic systems with coupled hysteresis and linear dynamics: application to piezoelectric-actuated nanopositioning stages. *IEEE/ASME Transactions on Mechatronics* 2019; 24(1): 326–337.

22. Xie M, Shakoor A, Li C, Sun D. Robust orientation control of multi-dof cell based on uncertainty and disturbance estimation. *International Journal of Robust and Nonlinear Control* 2019; 29(14): 4859–4871.

23. Jiang B, Karimi HR, Yang SC, Kao Y, Gao C. Takagi-Sugeno model-based reliable sliding mode control of descriptor systems with semi-Markov parameters: average dwell time approach. *IEEE Transactions on Systems, Man, and Cybernetics: Systems* 2019; 51(3): 1549–1558.

24. Stepanenko Y, Su CY. Variable structure control of robot manipulators with nonlinear sliding manifolds. *International Journal of Control* 1993; 58(2): 285–300.

25. Jiang B, Karimi HR, Kao Y, Gao C. Takagi-Sugeno model based event-triggered fuzzy sliding mode control of networked control systems with semi-Markovian switchings. *IEEE Transactions on Fuzzy Systems* 2019; 28: 1–1.

26. Chouza A, Barambones O, Calvo I, Velasco J. Sliding mode-based robust control for piezoelectric actuators with inverse dynamics estimation. *Energies* 2019; 12(5): 943.

27. Lee J, Chang PH, Jin, M. Adaptive integral sliding mode control with time-delay estimation for robot manipulators. *IEEE Transactions on Industrial Electronics* 2017; 64(8): 1–1.

28. Wang Y, Zhu K, Yan F, Chen B. Adaptive super-twisting nonsingular fast terminal sliding mode control for cable-driven manipulators using time-delay estimation. *Advances in Engineering Software* 2019; 128(FEB.): 113–124.

29. Wu Z, Jiang B, Kao Y. Finite-time $\mathcal{H}_\infty$ filtering for itô stochastic markovian jump systems with distributed time-varying delays based on optimisation algorithm. *IET Control Theory and Applications* 2019; 13(5): 702–710.

30. Levant A. Higher-order sliding modes, differentiation and output-feedback control. *International Journal of Control* 2003; 76(9–10): 924–941.

31. Lee J, Jin M, Kashiri N, Caldwell DG, Tsagarakis NG. Inversion-free force tracking control of piezoelectric actuators using fast finite-time integral terminal sliding-mode. *Mechatronics* 2019; 57: 39–50.

32. Sun XM, Du SL, Shi P, Wang W, Wang LD. Input-to-state stability for nonlinear systems with large delay periods based on switching techniques. *IEEE Transactions on Circuits & Systems I Regular Papers* 2014; 61(6): 1789–1800.

33. Jin M, Yi J, Chang PH, Choi C. High-accuracy trajectory tracking of industrial robot manipulators using time delay estimation and terminal sliding mode. In: *Conference of the IEEE Industrial Electronics Society*. IEEE. 2009.

34. Hsia TCS, Lasky TA, Guo Z. Robust independent joint controller design for industrial robot manipulators. *IEEE Transactions on Industrial Electronics* 1991; 38(1): 21–25.

35. Zhang J, Cheng M. A real time testing system for wind turbine controller with XPC target machine. *International Journal of Electrical Power & Energy Systems* 2015; 73(dec.): 132–140.

36. Wienholds E, Kloosterman W, Miska E, Alvarez-Saavedra E, Berezikov E, Bruijn ED et al. MicroRNA expression in zebrafish embryonic development. *Science*, 2005; 309(5732): 310–311.

7 Micro-Force Tracking Control of Cell Puncture Mechanism Based on Time-Delay Estimation Technology

7.1 INTRODUCTION

Previous research results have built a nonlinear robust controller based on the dynamic model of the cell puncture mechanism to realize the precise motion control of the cell puncture mechanism. Theoretically, in the position control mode, the glass microneedle can continuously feed and push the cell membrane to deform and, finally, achieve the goal of puncture. However, such puncture technology has two unavoidable adverse consequences because the cell is a biological unit with extremely fragile vitality:

(1) The large-scale and continuous deformation of the cell embryo will destroy the biological structure inside the cell, thereby affecting the subsequent growth and development of the cell;
(2) The cell puncturing technology under the position control model cannot accurately grasp the critical moment when the cell membrane is punctured.

Therefore, it is needed to urgently explore and overcome the defects of cell puncture technology under the simple position control model. Although various scientists are committed to using machine vision technology to make up for the lack of position control [1], great efforts have also been made. However, its limitations are also evident: the cell membrane is a white transparent biofilm [2], and its texture is similar to that of glass microneedles. Therefore, extracting the shape of glass microneedles in the area where the cell membrane is deformed is difficult. In addition, it is extremely difficult for machine vision to detect 5-micron glass microneedles and realize real-time feedback [3]. Therefore, the development prospect of machine vision in cell puncture operation is not optimistic.

The two shortcomings of position control in cell puncture technology can be solved by micro-force tracking. Different from the position control, the micro-force tracking control is often used to monitor the micro-force signal during cell puncture, from the moment when the glass microneedle contacts the cell membrane to the

DOI: 10.1201/9781003294030-7

whole process when the glass microneedle penetrates the cell membrane. The cell puncturing technology can penetrate the cell membrane that undergoes elastic deformation with an incrementally increasing cell puncturing force by tracking the force signal based on the micro-force tracking control. More importantly, at the moment when the cell membrane is penetrated, the controller can quickly sense the critical moment when the cell membrane is punctured because of the transient mutation of the micro-force signal and stop moving.

In the intelligent control theory, the force control methods mainly include impedance control and force tracking control. The following is a comparative analysis:

Hogan proposed the famous "impedance control" theory in 1984 [4], and it is widely used in force tracking control or compliance control [5]. The key idea of impedance control is to change the mechanical impedance of the robot to obtain the desired impedance characteristics, integrate the position control in free space, and the compliance control in contact space into one system [6], as well as coordinate the compliance control of force and position with a single controller.

In addition, impedance control can be divided into impedance control (force-based impedance control) and admittance control (position-based impedance control) in terms of its implementation mode [7]. For admittance control, lower desired stiffness and damping will produce higher feedback gain, thereby leading to instability when interacting with a high stiffness environment. On the contrary, in impedance control, high feedback gain is the result of high desired stiffness and damping; hence, it is difficult for the system to provide large desired stiffness, thereby resulting in poor accuracy of the robot in a low stiffness environment and free motion. Herein, impedance control and admittance control systems have complementary advantages and disadvantages. Moreover, neither of them can provide the best performance for a wide range of tasks, from precise movement in free space to stable dynamic interaction with rigid environments.

Given that the impedance control tracks the force signal by controlling the position signal, the tracking accuracy of this "indirect" control method for the force signal is far inferior to the direct tracking of the force signal. Therefore, it will be a direct and effective force tracking method to take the resistance strain micro-force sensor as the sensor of micro-force collection and build a robust controller with the micro-force signal as the tracking target.

Although, based on PEAs, various scientists have conducted in-depth and comprehensive research on the position control of PEAs. However, the micro-force tracking control is rarely reported. Simultaneously, the hysteretic nonlinear effect of PEAs also brings a remarkable effect on the micro-force tracking control big problem. Meanwhile, in previous studies on force tracking control, no reliable solutions have been proposed for the estimation of unmodeled dynamics or unknown terms in the model. For example, Adibi et al. [7] realized force tracking control based on the sliding mode control technology of force feedback. Moreover, they built a disturbance observer (DOB) based on the secondary derivation of force signal in the controller to realize real-time estimation of unknown items. However, this results in the accumulation and expansion of errors in the force signal. In Chapter 5 of this book, i.e. Motion Control of Cell Puncture Mechanism based on FONTSM, TDE is used to realize the real-time estimation and online compensation of unknown items, which reflects the effectiveness of TDE technology. Furthermore, TDE technology will still be used to estimate and compensate for the unknown items in the design of the micro-force tracking controller.

This chapter will focus on building the micro-force tracking control of the cell puncture mechanism, which will lay the foundation for the subsequent force-position mixing control. First, based on the hysteresis dynamics model of the cell puncture mechanism, a micro-force tracking controller is designed, and the TDE technology is used to estimate the unknown items; second, the PID servo controller is used to realize the fast response of the transient force and eliminate the steady-state error. Subsequently, the stability proof of the closed-loop system is completed by using the bounded input bounded output (BIBO) stability theory. Finally, the effectiveness of the proposed controller is demonstrated by computer simulation and hardware-in-the-loop simulation experiments.

7.2 KINETIC MODEL OF CELL PUNCTURE PROCESS

Herein, the equivalent kinetic model of the cell puncturing mechanism during the puncturing process is established: mass-spring-damping system, as shown in Figure 7.1. f_a, τ_d, and f_e are the output forces of PEAs, the external disturbance of the system, and the cell puncture force, respectively.

Using Newton's law of motion and combining the Bouc–Wen model, the complete kinetic model of the cell puncturing mechanism including hysteresis during the puncturing process is as follows:

$$m\ddot{x} + b\dot{x} + kx = f_a - \tau_d - f_e = k(du - h) - \tau_d - f_e \tag{7.1}$$

To simplify the above formula, the voltage coefficient is defined as $\zeta = kd$. Generalized Hysteresis Variable $\hbar = kh$.

Then, the above formula will be rewritten as:

$$m\ddot{x} + b\dot{x} + kx + \tau_d = \zeta u - \hbar - f_e \tag{7.2}$$

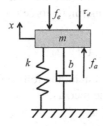

FIGURE 7.1   Equivalent kinetic model of cell puncture mechanism during cell puncture.

7.3 DESIGN OF MICRO-FORCE TRACKING CONTROLLER

7.3.1 CONTROLLER DESIGN

Define the force error between the desired force and the cell puncture force as:

$$\delta_f = f_d - f_e \tag{7.3}$$

The goal of micro-force tracking control is to enable the cell-piercing mechanism to achieve precise tracking of the desired force, that is, despite the presence of disturbances or hysteretic nonlinear effects in the system.

In addition, the control of the cell puncture mechanism is realized by inputting the voltage, and the complete kinetic model of the cell puncture mechanism including the hysteresis link in the puncture process is rearranged, and the result is obtained.

$$u = \zeta^{-1}\left(m\ddot{x} + b\dot{x} + kx\right) + \zeta^{-1}\left(\tau_d + \hbar\right) + \zeta^{-1}f_e \tag{7.4}$$

$$\eta = \zeta^{-1}\left(\tau_d + \hbar\right) \tag{7.5}$$

where η includes lumped uncertainty terms including disturbance and hysteresis non-linear effects, which cannot be obtained by mathematical analysis.

Transform (7.4) into

$$u = \zeta^{-1}\left(m\ddot{x} + b\dot{x} + kx\right) + \eta + \zeta^{-1}f_e \tag{7.6}$$

To achieve accurate micro-force tracking, the control law is defined as:

$$u_f = u_{fTDE} + u_{auxiliary} \tag{7.7}$$

In the micro-force tracking control law, the first term on the right side of the equation is used to estimate and compensate for the unknown term; whereas, the second term is the auxiliary servo control term, which is mainly used to realize the expected error dynamic equation and construct the transient response of the closed-loop system. Therefore, for the following:

$$u_{fTDE} = \hat{\eta} \tag{7.8}$$

TDE technology will be applied to realize online estimation and real-time compensation of unknown items during micro-force tracking.

$$u_{fTDE} = \hat{\eta} = u_{(t-L)} - \zeta^{-1}\left(m\ddot{x}_{(t-L)} + b\dot{x}_{(t-L)} + kx_{(t-L)} + f_{e(t-L)}\right) \tag{7.9}$$

For the micro-force tracking system, the following first-order PID-type expected force error dynamic equation is defined as:

$$\delta_f + K_i \int \delta_f d\tau + K_d \dot{\delta}_f = 0 \tag{7.10}$$

This simple PID-type error dynamic equation will help to simplify the control algorithm. The auxiliary servo control item is designed as follows:

$$u_{\text{auxiliary}} = \zeta^{-1}\left(m\ddot{x} + b\dot{x} + kx + f_d + K_i \int \delta_f d\tau + K_d \dot{\delta} f\right) \tag{7.11}$$

To sum up, the micro-force tracking control law will be defined as:

$$u_f = u_{(t-L)} - \underbrace{\zeta^{-1}\left(m\ddot{x}_{(t-L)} + b\dot{x}_{(t-L)} + kx_{(t-L)} + f_{e(t-L)}\right)}_{u_{\text{TDE}}}$$
$$+ \underbrace{\zeta^{-1}\left(m\ddot{x} + b\dot{x} + kx + f_d + K_i \int \delta_f d\tau + K_d \dot{\delta}_f\right)}_{u_{\text{auxiliary}}} \tag{7.12}$$

Given that the proposed micro-force tracking controller mainly includes the TDE term and the PID auxiliary servo control term, the proposed controller is abbreviated as the TDE-PID controller.

Based on the designed control law, the real-time estimation of the full state is realized by RED technology; the auxiliary servo control item realizes the precise control of the micro-force tracking, as well as the control system block diagram is drawn, as shown in Figure 7.2.

7.3.2 STABILITY ANALYSIS

The stability analysis of the controller is given below.

Substitute control law (7.12) into (7.6), thus we get:

$$\zeta(\eta - \hat{\eta}) = \delta_f + K_i \int \delta_f d\tau + K_d \dot{\delta}_f = \xi \tag{7.13}$$

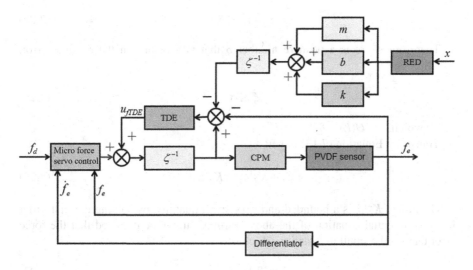

FIGURE 7.2 System block diagram of the micro-force tracking control.

Eq. (7.13) shows the residual error of the closed-loop system dynamics equation after applying the TDE technique. It is necessary to prove the boundedness of this residual error to prove the stability of the proposed controller.

Rewrite the control law (7.12) as:

$$u_f = \hat{\eta} + u_{\text{auxiliary}} = \hat{\eta} + \zeta^{-1}\left(m\ddot{x} + b\dot{x} + kx + f_d + K_i \int \delta_f d\tau + K_d \dot{\delta}_f\right) \qquad (7.14)$$

Thus combining Equations (7.13) and (7.14), we get:

$$u_f + \zeta^{-1}\xi = u_{\text{auxiliary}} + \eta \qquad (7.15)$$

Arranging and transforming Eq. (7.15) several times, thus we get:

$$\xi = \zeta u_{\text{auxiliary}} + \zeta \eta - \zeta u_f \qquad (7.16)$$

$$\xi = -\zeta\hat{\eta} + \tau_d - \hbar \qquad (7.17)$$

$$\xi = (\tau_d - \hat{\tau}_d) - (\hbar - \hat{\hbar}) \qquad (7.18)$$

From Eq. (7.18), although the generalized hysteresis variable is not a smooth curve, it is a Lipschitz continuous curve [8]. From the property of Lipschitz continuity, it can be inferred that there is a sufficiently small delay constant L such that:

$$\left|\hbar - \hat{\hbar}\right| = \left|\hbar - \hbar_{(t-L)}\right| \le O(L) \qquad (7.19)$$

The external disturbance of the system τ_d is a discontinuous but bounded variable. Therefore, a positive number exists, which satisfies:

$$\left|\tau_d - \hat{\tau}_d\right| \le \delta_d \qquad (7.20)$$

Therefore, there is a positive number σ that is a bound on the residual error, namely:

$$\left|\xi\right| \le \sigma \qquad (7.21)$$

Of which, $\sigma = O(L) + \delta_d$

Transform Equation (7.13) to get:

$$\dot{\delta}_f + K_d^{-1}\delta_f = K_d^{-1}\xi \qquad (7.22)$$

Given that $K_d^{-1}\delta_f$ is a bounded and very small quantity, by solving the first-order linear differential equation of the above formula, it can be obtained that the force error tends to be small in quantity, which is:

$$\lim_{t\to\infty}\left|\delta_f\right| \le K_d^{-1}\sigma \qquad (7.23)$$

Therefore, the closed-loop system was constructed using the proposed controller for micro-force tracking control that conforms to the BIBO stability [9].

The stability analysis of the closed-loop system is thus far, and then the computer simulation experiment is carried out.

7.4 SIMULATION EXPERIMENT OF MICRO-FORCE TRACKING CONTROL

In this section, a computer comparative simulation experiment will be performed on the proposed controller to investigate the performance of the proposed controller.

7.4.1 PID CONTROLLER FOR MICRO-FORCE TRACKING

Given the deep recognition and wide application of the PID controller [10] in industry and academia, a comparison with the PID controller will be made to verify the performance of the proposed controller.

Based on the force error formula $\delta_f = f_d - f_e$ between the desired force and the cell puncture force, the PID controller based on micro-force tracking is constructed as follows:

$$u_{\mathrm{PID}} = K_P\delta_f + K_I\int\delta_f d\tau + K_D\dot{\delta}_f \tag{7.24}$$

7.4.2 SIMULATION RESULTS

The performance index of the proposed controller in an ideal environment is investigated through computer simulation experiments; the potential of the control is tapped; the debugging progress is accelerated; and the success rate of physical experiments is improved. Herein, the environment of the computer simulation experiment is similar to the previous setting. In addition, the computer is Windows 7 operating system, 3.60 GHz CPU with 8G memory, the simulation software used is MATLAB/Simulink 2016b, and the time step is 0.1 ms.

Taking the tracking accuracy of the force signal as the target of the controller parameter debugging, the proposed controller has a very high debugging efficiency and strong applicability because of the simple structure of the proposed control and few parameters to be debugged. After debugging, the parameters used are $K_i = 1200$, $K_d = 2500$.

First, let the controller track an ideal sinusoidal signal as the desired force signal to examine the performance of the controller, whose expression is as follows:

$y = \left(100 + 100 \times \sin\left(\pi t - \pi/2\right)\right)$ mN. The curve is shown in Figure 7.3.

The control laws of the two controllers (TDE-PID controller and PID controller) are obtained by using Formulas (7.12) and (7.24), respectively, and the output curves of the controllers are shown in the figure. The proposed TDE-PID controller has a smaller amplitude control law curve for tracking the sinusoidal force curve, as demonstrated in Figure 7.4; therefore, in terms of energy consumption, the TDE-PID

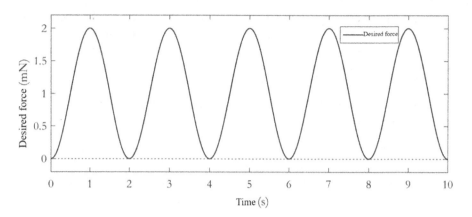

FIGURE 7.3   Sinusoidal force curve for tracking.

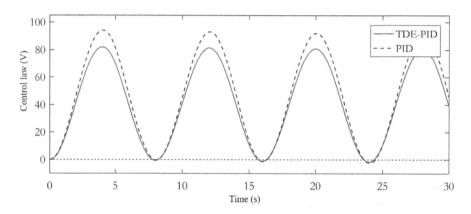

FIGURE 7.4   Control law curves obtained by the two controllers when tracking the sinusoidal force curve.

control is more energy-efficient. In addition, the PID controller has the problem of excessive energy loss when performing micro-force tracking. Under the premise that the control law output curves are smooth and without chattering, it is necessary to analyze the micro-force tracking accuracy of the controllers to explore the energy consumption performance of the two controllers.

As illustrated in Figure 7.5, both controllers achieve accurate tracking of the sinusoidal force curve, and the proposed TDE-PID controller has a higher micro-force tracking accuracy than the traditional PID controller. In general, the more accurate the dynamic model of the plant is integrated into the controller, the better the quality of the controller will be. Real-time estimation and online compensation of unknown items are realized by using TDE technology, and some information in auxiliary servo control items are obtained by combining laser displacement sensors. Hence, the controller can directly realize the output of most control laws through estimation and acquisition to achieve more high control accuracy. Moreover, further suppression of

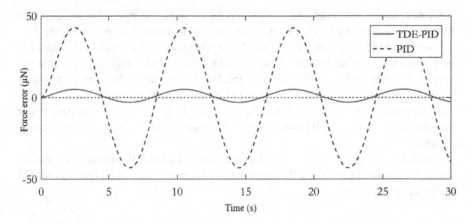

FIGURE 7.5 The force error curves obtained by the two controllers when tracking the sinusoidal force curve.

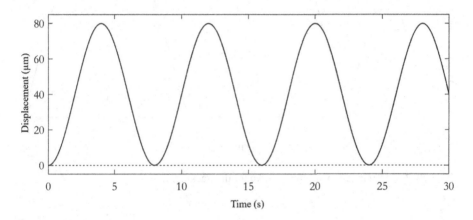

FIGURE 7.6 The displacement curve generated by the TDE-PID controller when tracking the sinusoidal force curve.

errors is achieved through the PID control link in the auxiliary servo control item. In comparison, for a PID controller not based on model design, its error adjustment all depends on the adjustment of the controller, and its control accuracy is inferior to that of the TDE-PID controller-based on model design.

The displacement curve of the cell puncturing mechanism when it is tracked by a micro-force under the action of the TDE-PID controller is similar to a sine curve, as shown in Figure 7.6.

In the design process of the TDE-PID controller, it is assumed that the first and second derivatives of the displacement signals in the system exist and are continuous; simultaneously, the first derivative of the micro-force signal exists and is continuous. However, the tracked curve may be continuous but non-differentiable in practical engineering applications, which will pose difficulties to the performance of the controller. In the next computer simulation experiment, the triangular wave micro-force curve will be used as the tracking object, and its curve is shown in Figure 7.7.

Both controllers can accurately track the triangular wave force curve, and the performance is similar to that of the sinusoidal force curve; this indicates that the proposed controller has stable performance. Figure 7.8 shows the energy consumption comparison of the control law; meanwhile, Figure 7.9 shows the tracking accuracy comparison. It can be seen that the proposed TDE-PID controllers have better performance, not only with lower energy consumption but also with higher tracking accuracy.

As shown in Figure 7.10, the displacement curve of the cell puncture mechanism under the action of the TDE-PID controller for the micro-force tracking is similar to the triangle wave force curve.

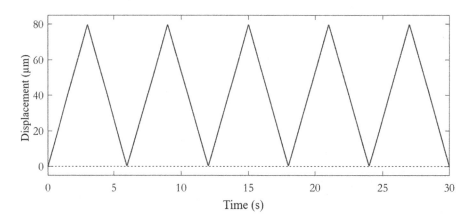

FIGURE 7.7   Triangular wave force curve for the tracking.

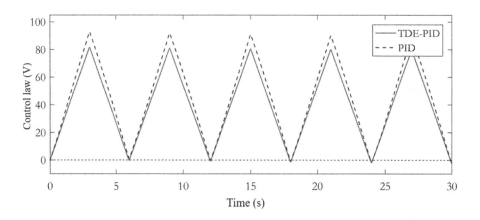

FIGURE 7.8   The control law curves obtained by the two controllers when tracking the triangular wave force curve.

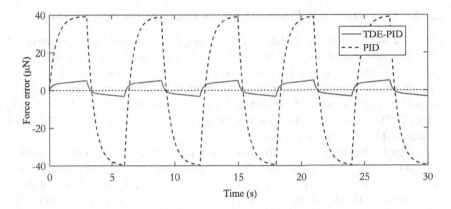

FIGURE 7.9 The force error curves obtained by the two controllers when tracking the sinusoidal force curve.

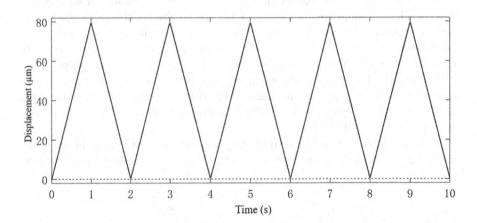

FIGURE 7.10 The displacement curve generated by the TDE-PID controller when tracking the triangular wave force curve.

7.5 HARDWARE-IN-THE-LOOP SIMULATION EXPERIMENT OF TDE-PID CONTROLLER

In addition, the proposed TDE-PID micro-force control will be experimentally studied through hardware-in-the-loop simulation experiments to investigate the performance of the proposed control in realizing micro-force tracking.

The micro-force acquisition sensor is the key component in this experimental system to realize force signal acquisition and feedback. Given the lack of micro-force sensors on the market that meet the requirements, the development process is given below.

7.5.1 DEVELOPMENT OF RESISTANCE STRAIN GAUGE MICRO-FORCE SENSOR

The body structure of the resistance strain gauge micro-force sensor is designed using the principle of a compliant mechanism, and an arc hinge with high swing precision is introduced as a rotating joint. Therefore, the overall structure presents a parallelogram layout to constrain the DOF of the detection part and effectively control the global rigidity of the body structure. Its 3D structure model is shown in Figure 7.11(a), and its installation schematic diagram is shown in Figure 7.11(b), and the material is 7075 aviation aluminum.

Similar to the structure optimization design process of the cell puncture mechanism in Chapter 2, herein, the finite element optimization design and strength check of the body structure of the resistance strain gauge micro-force sensor were carried out using the ANSYS Workbench development platform. In addition, under the premise of ensuring the strength and natural frequency, when the detection part is subjected to a load of 100 mN, the displacement of the detection part is less than 15 μm. The optimized design aims to make the thickness of the arc hinge the thinnest, which is beneficial to the acquisition of the sensor.

The stress–strain nephogram of the bulk structure obtained from ANSYS Workbench is demonstrated in Figure 7.12, herein, the strain gauges need to be arranged on the front and back of the arc hinge. Moreover, the front and back of the arc hinge are the concentrated areas where strain occurs. Paste the strain gauges in this plane area, which has the effect of "good steel is used at the cutting edge".

The HU-101B-350 semiconductor strain gauge is used as the strain gauge for collecting micro-force signals. It has the characteristics of small substrate size, high sensitivity, and low-temperature sensitivity, as well as accurate collection. Its parameters are shown in Table 7.1.

There are four semiconductor strain gauges attached to the front and back of the arc hinge, and the acquisition circuit is built based on the principle of the Wheatstone

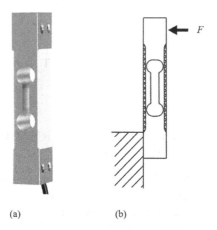

(a)   (b)

FIGURE 7.11 The body structure of the resistance strain gauge micro-force sensor; (a) Outline structure of the body structure, (b) Schematic diagram of the installation of the body structure.

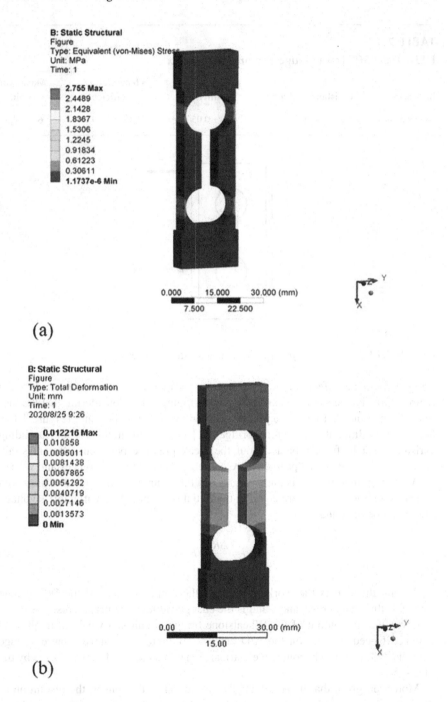

FIGURE 7.12 The stress–strain cloud diagram of the body structure obtained through ANSYS Workbench; (a) Stress nephogram of the bulk structure, (b) Deformation cloud map of ontology structure.

TABLE 7.1
HU-101B-350 Strain Gauge Performance Index

Base area	Resistance value	Sensitivity	Temperature coefficient	Maximum strain
$5mm \times 3mm$	350Ω	130 ± 0.05	$\leq 0.15 \% /^{\circ}C$	$6 \times 10^3 \mu\varepsilon$

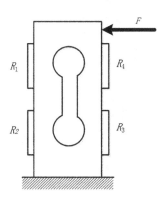

FIGURE 7.13 Schematic layout of semiconductor strain gauges.

bridge, which can effectively compensate for the measurement error caused by the temperature change to the sensor, as shown in Figure 7.13. Considering that the fabricated sensor needs to achieve precise measurement of tiny pressure (Figure 7.14), hence the traditional Wheatstone bridge [11] is partially modified, and a sliding varistor is set to further balance out the weak pressure between different strain gauges. Bridge imbalance problems are caused by resistance differences and wires.

Assuming that the resistance values of the four branches of the improved Wheatstone bridge circuit are equal in the initial state, based on Kirchhoff's voltage law, it can be obtained:

$$U_{ou} = \frac{\Delta R_i}{R_i} \times E \qquad (7.25)$$

Among them, E is the working voltage, the output voltage of the Wheatstone bridge, is the branch resistance, and is the change value of the branch resistance.

Given that the output of the Wheatstone bridge circuit is a weak millivolt level voltage, it needs to rely on the A/D conversion circuit to achieve a large voltage amplification factor to become an electrical signal that can be effectively used by the xPC system.

Moreover, given that noise affects the resolution of the sensor, the resolution of the sensor is increased by a factor of 10, and the resolution of the circuit is at least. Chipsea's CS1237-SO high-precision 24-bit A/D conversion chip is chosen, which can achieve the ideal resolution, and the magnification of the chip is set to 128 times. The maximum range, the number of divisions, and the minimum division value are set to 200 mN, 2,000, and 0.1 mN, respectively.

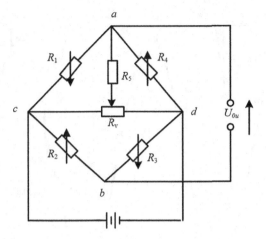

FIGURE 7.14 Modified Wheatstone bridge circuit.

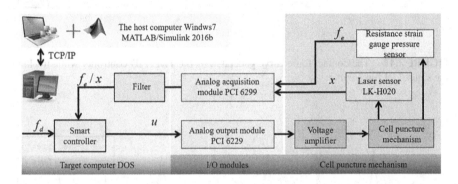

FIGURE 7.15 Flow chart of the control system of the micro-force tracking experiment.

The developed resistance strain gauge micro-force sensor is calibrated with standard weights and used for the following micro-force tracking experiments.

7.5.2 Micro-Force Tracking Experiment

The flow chart of the hardware-in-the-loop simulation experiment is designed based on the proposed TDE-PID controller as shown in Figure 7.15. Different from the previous displacement tracking experiments, in this controller, it is necessary to realize the real-time acquisition and feedback of the displacement and contact force of the cell puncture mechanism simultaneously.

First, let the controller track the ideal sinusoidal signal as the desired force signal. The cell puncture force collected by the sensor can accurately track the desired force, as shown in Figure 7.16. Herein, the sensor overshoots at this moment, however, it can reach stability within 0.2 s and achieve the fitting with the expected force signal to quickly reach the puncture force of 50 mN.

Furthermore, the contact force curve generated by the TDE-PID controller when tracking the triangular wave force curve is shown in Figure 7.17. Although the

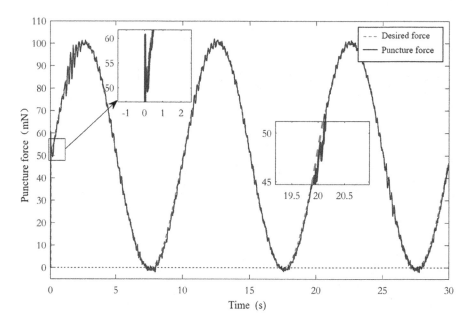

FIGURE 7.16 The contact force curve generated by the TDE-PID controller while tracking the sinusoidal force curve.

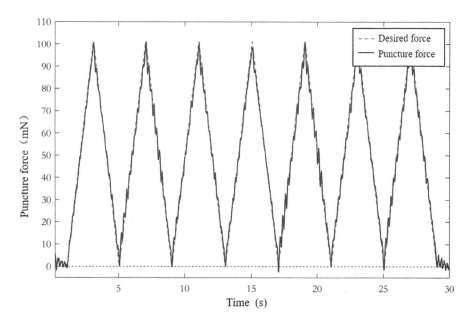

FIGURE 7.17 The contact force curve generated by the TDE-PID controller when tracking the triangular wave force curve.

triangular wave force curve is non-differentiable, both simulation and physical experiments prove that the proposed TDE-PID controller can accurately track the triangular wave force curve.

7.6 CONCLUSION

In this chapter, for the hysteresis dynamic models of the cell puncture mechanism, the TDE technology is used to realize real-time estimation of unknown items and online compensation, simultaneously, PID servo control items are used to achieve fast convergence and suppression of errors. The main contributions of this chapter are as follows:

First, In the controller's design, the stability proof of the controller is completed using the BIBO stability theory. The computer simulation experiment proves that the expected force can be accurately tracked using the continuous differentiable sinusoidal force signal, meanwhile, the continuous but non-differentiable triangular wave force signal is the expected value.

Second, in the physical experiment, a resistance strain gauge micro-force sensor was developed to realize the precise acquisition of the micro-force signal. The mechanical design, structure optimization, and strength check of the sensor have been completed, and a high-precision 24-bit A/D conversion module has been developed to realize the acquisition and amplification of the weak signal at the millivolt level.

Finally, it is verified through the hardware-in-the-loop simulation experiment that the proposed TDE-PID control can effectively realize the precise tracking of the micro-force signal in the physical implementation. The micro-force tracking control of the cell puncture mechanism constructed herein lays the foundation for the subsequent force-position mixing control in chapter 8.

REFERENCES

1. Thryft AR. Machine-vision & inspection test report: lenses in solar, flat-panel inspection. *Test & Measurement World the Magazine for Quality in Electronics* 2011.
2. Shi Y, Massagué J. Mechanisms of TGF-β signaling from cell membrane to the nucleus - sciencedirect. *Cell* 2003; 113(6): 685–700.
3. Ordaz MA, Lush GB, Tobin KW. Machine vision for solar cell characterization. *Proceedings of SPIE - The International Society for Optical Engineering* 2000; 3966: 238–248.
4. Hogan N. Impedance control: an approach to manipulation: part iii—applications. *Journal of Dynamic Systems Measurement & Control* 1985; 107(1): 17–24.
5. Hogan N. Impedance control - an approach to manipulation. i - theory. ii - implementation. iii - applications. *Journal of Dynamic Systems Measurement and Control* 1985; 107: 1–24.
6. Selen L, Franklin DW, Wolpert DM. Impedance control reduces instability that arises from motor noise. *Journal of Neuroscience* 2009; 29(40): 12606–12616.
7. Abidi K, Sabanovic A, Yesilyurt S. Sliding-mode based force control of a piezoelectric actuator. In: *IEEE International Conference on Mechatronics.* IEEE. 2004.
8. Logemann H, Ryan EP, Shvartsman I. A class of differential-delay systems with hysteresis: asymptotic behaviour of solutions. *Nonlinear Analysis Theory Methods & Applications* 2008; 69(1): 363–391.

9. Lee EB, Luo JC. On "evaluating the bounded-input-bounded-output stability integral for second-order systems". *IEEE Transactions on Automatic Control* 2000; 45(2): 311–312.

10. Lee Y, Park S, Lee M, Brosilow C. PID controller tuning for desired closed-loop responses for SI/SO systems. *AIChE Journal* 1998; 44(1): 106–115.

11. Cowles VE, Condon RE, Schulte WJ, Woods JH, Sillin LF. A quarter wheatstone bridge strain gage force transducer for recording gut motility. *The American Journal of Digestive Diseases* 1978; 23(10): 936–939.

8 Hybrid Control Strategy of Force and Position for Cell Puncture Based on Adaptive Smooth Switching

8.1 INTRODUCTION

In force position control, the operation space is divided into position control and force control subspaces. The great advantage of impedance/admittance control in the wide application of force level control depends on its integration of force and position controls into a unified control system, as well as the switch between force control and position control subspaces. Although impedance/admittance control can achieve a smooth transition of the switching process, it depends on the precise tuning of the impedance parameters after careful investigation of the operating environment. Literatures [1, 2] proposed a switching algorithm based on fuzzy logic to reduce the difficulty of impedance parameter setting. However, once the logical rules are determined, they cannot be automatically updated in different environments and applications.

In the complex application of biological cell puncture, their physical characteristics are changing with growth and development because biological cells have life characteristics. Taking zebrafish embryos as an example, the time from birth to hatching into small fish is 24 hours; and the physical parameters of the cell membrane were constantly changing during the 24-hour course. As shown in Figure 8.1, the young's modulus of the four development stages of zebrafish embryos, including blastocyst (b), gastrula (g), pharyngeal stage embryo (P), and preincubation embryo (PH), is different [3]. Moreover, the experiment found that the stiffness coefficient of different zebrafish embryos at the same stage is different [4]. Therefore, the robustness of the force controller must be strong enough to overcome the unknown stiffness during cell puncture.

The cell puncture mechanism realizes the position control of the cell puncture mechanism in the free space and the force control with the cell in the force control subspace. Meanwhile, in the subspace of force control, impedance/admittance control indirectly controls the force signal by controlling the position signal [5]. The robustness of this indirect control is not as strong as that of the force signal's direct control [6].

DOI: 10.1201/9781003294030-8

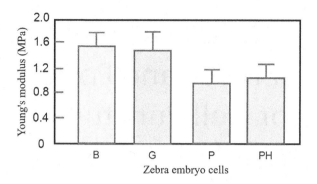

FIGURE 8.1   Young's modulus of zebrafish embryonic cells at different development stages.

However, the difficulty of force position hybrid control is how to realize the "automatic" switching of force control and position control subspaces and ensure that the switching process is transitioning smoothly without effect and mutation. Therefore, few teams have conducted in-depth and comprehensive research in this area.

In Chapters 5 and 6 of this book, the position control and micro-force tracking control of the cell puncture mechanism has been completed, respectively, which lays a foundation for the force position hybrid control and compliance switching of cell puncture in this chapter. Therefore, this chapter's key content will focus on the design and implementation of a smooth handover algorithm. First, from the point of view of easy engineering application, a switching algorithm of coupling force and position controls is proposed, which brings the motion control of cell puncture mechanism based on fractional nonsingular terminal sliding mode and the micro-force tracking control of cell puncture into a unified control system. Subsequently, taking zebrafish embryo cells as the verification object, the force position hybrid control and compliance switching experiments of cell puncture were completed to verify the superiority and progressiveness of the proposed controller design.

8.2   CELL PUNCTURE STRATEGY

During the manual puncture of zebrafish embryos [7], as shown in Figure 8.2, zebrafish embryo cells need to be arranged in a whole row on a mold, and then the glass microneedles are placed close to the surface of zebrafish embryos. Afterward, the operator drives the glass microneedles to undergo a fixed stroke displacement, thus quickly penetrating the cell membrane and reaching the inside of the embryonic tissue. The external substance will then be injected into the tissue structure. Finally, the glass microneedle returns to its original position. Thus, one cycle of cell puncture and injection was completed. In a biological experiment, both drug development and disease tracking require a high-throughput sample number. Therefore, the injection of zebrafish embryos will be a huge amount of work [8].

The survival rate of zebrafish embryos is greatly reduced because it is impossible to construct a reasonable control strategy to reduce the elastic deformation of the cell

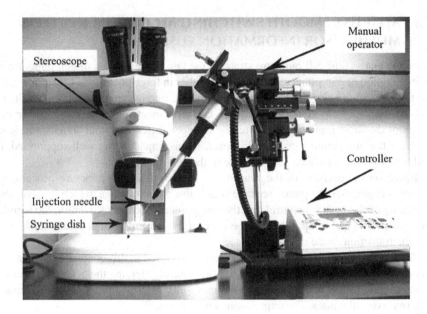

FIGURE 8.2   Manual puncture equipment for zebrafish embryos.

membrane during manual injection [9]. Second, it is easy to cause fatigue and misoperation to complete a fine operation process manually and repeatedly.

In addition, in developing the automatic puncture strategy based on the cell puncture mechanism, the manual puncture process will be used for reference. Considering the puncture task of zebrafish embryonic cells, it is divided into four stages: approach stage, puncture stage, positioning stage, and exit stage. Therefore, certain operation strategies will be designed in these four stages to improve the survival rate of cell puncture and overcome the disadvantages of manual operation. The operation strategy is as follows:

(a) In the approach stage, the glass microneedles approach the surface of zebrafish embryonic cells at a constant speed driven by the cell puncture mechanism. Herein, the cell puncture mechanism works in the position control subspace;

(b) In the puncture stage, the contact force of the glass microneedle will gradually increase after it contacts the surface of the cell membrane. When the contact force reaches a threshold f_{p2f}, at this time, the cell puncture mechanism will enter the force control subspace from the position control subspace; when the cell puncture is completed, the cell puncture mechanism will enter the position control subspace from the force control subspace;

(c) Meanwhile, in the positioning and the withdrawal phases, the cell puncture mechanism is in the position control subspace to complete accurate motion control.

8.3 ADAPTIVE SMOOTH SWITCHING ALGORITHM BASED ON MULTISENSOR INFORMATION FUSION

From the analysis of cell puncture strategy, the key to the success of cell puncture is the realization of compliance switching algorithm in force position hybrid control. In the force position hybrid control, the input and the outputs are the excitation voltage and the displacement signal or the force signal, respectively. Therefore, under the action of the same input signal, switching between different output signals will lead to discontinuous output signals and impact. If this impact is not well suppressed, it will bring additional effects and damage to the cell membrane.

Based on the division of the four stages of the cell micro puncture process, as shown in Figure 8.3, the puncture stage is the most complex stage, which involves the switching of the position control and the force control subspaces. Herein, the weight factor is introduced to characterize the state of the controller in different subspaces and the transition phase in two subspaces.

An adaptive smooth switching algorithm is proposed in this chapter to realize the smooth switching of force position hybrid control. Herein, the adaptive smooth switching algorithm is used to make the controller realize smooth switching under different control modes. The expression is as follows:

$$u_{\text{synthesize}} = u_{\text{p}}\hat{\rho} + u_{\text{f}}\left(1 - \hat{\rho}\right) \tag{8.1}$$

u_p represents the motion controller based on fractional nonsingular terminal sliding mode designed

$$u_{\text{p}} = \underbrace{\frac{1}{kd}\left\{m\left[\ddot{x}_d - kD^\lambda\left(\text{sig}(e)^\alpha\right)\right] + b\dot{x} + kx - m\left[k_1 s + k_2\text{sig}(s)^\beta\right]\right\}}_{FONTSM}$$

$$\underbrace{-\frac{m}{kd}\left\{\ddot{x}_{(t-L)} - \frac{kd}{m}u_{(t-L)} + \frac{b}{m}\dot{x}_{(t-L)} + \frac{k}{m}x_{(t-L)}\right\}}_{TDE} \tag{8.2}$$

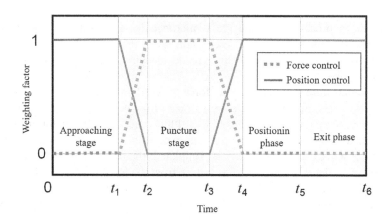

FIGURE 8.3 Change strategy of weight factor in force level hybrid control.

u_f is the micro-force tracking controller designed as:

$$u_f = \underbrace{u_{(t-L)} - \zeta^{-1}\left(m\ddot{x}_{(t-L)} + b\dot{x}_{(t-L)} + kx_{(t-L)} + f_{e(t-L)}\right)}_{u_{TDE}}$$

$$+ \underbrace{\zeta^{-1}\left(m\ddot{x} + b\dot{x} + kx + f_d + K_i\int\delta_f dt + K_d\dot{\delta}_f\right)}_{u_{auxiliary}}$$

(8.3)

ρ is the weight factor, obtained by an adaptive algorithm, and its formula is as follows:

$$\dot{\hat{\rho}} = k_f\left(f_e - f_{p2f}\right) \quad 0 \le \hat{\rho} \le 1$$

(8.4)

where k_f and f_{p2f} denote the growth coefficient and the threshold value for a puncture, respectively.

The advantage of using the adaptive algorithm to obtain the weight factor ρ is that the change of the weight factor ρ is continuous, and it can adapt to the changing trend of the contact force f_e and automatically adjust. For example, under the premise that $f_e > f_{p2f}$ and $\dot{f}_e > 0$, when the contact force f_e increases at a rapid speed, the weight factor ρ naturally also rapidly increases, thereby indicating that the adaptive smooth switching algorithm needs to switch the cell puncture mechanism from the position control subspace to the force control subspace at a faster speed and vice versa. Therefore, the weight factor ρ obtained by the adaptive algorithm makes the handover process smoother than the constant weight factor.

As shown in Figures 8.4 and 8.5, the change curve of weight factor ρ with contact force f_e is simulated. Herein, the contact force is amplified to improve the intuitiveness of the comparison. In general, when the glass microneedles approach the cells at a predetermined speed, the contact force presents a ramp-up trend while it was simulated that the contact force sharply decreased at the moment when the cell was penetrated, approximately 3 s. Therefore, a Gaussian noise signal of a certain

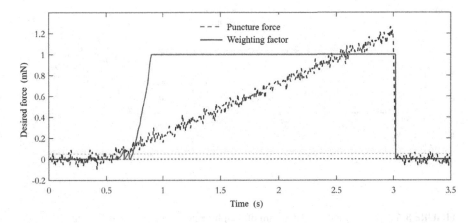

FIGURE 8.4 Simulation curve of change of weight factor ρ with contact force f_e.

frequency is applied to the contact force to more realistically display the data state of the contact force in the real environment.

In an ideal case, the contact force f_e exceeds the puncture threshold f_{p2f}, which is the time when the puncture starts. In the simulation, it is assumed that the puncture threshold p2ff is 0.05 N. However, although the contact force f_e gradually increases at the stage when the puncture process begins to occur, the contact force exceeds the puncture threshold at some times (e.g., at the time of 0.35 s, 0.37 s, and 0.48 s) because of the noise, however, it falls back rapidly, as shown in Figure 8.5. Although the weight factor ρ obtained by the adaptive algorithm slightly fluctuates at the above-mentioned corresponding time, it rapidly drops afterward, thus showing that the adaptive smooth switching algorithm of weight factor can actively overcome the influence of noise on the puncture process.

The puncture force showed an obvious increasing trend during the period from 0.6 s to 0.7 s. However, 13 the contact force was lower than the puncture threshold at a few moments because of the influence of noise, which was generally considered to be the time when the glass microneedle contacted the cell membrane and began to puncture. There is a fuzzy boundary in the process of puncture, and it is difficult to scale with an accurate time. Fortunately, the weight factor ρ obtained by the adaptive smooth switching algorithm can automatically adapt to the change of contact force, as well as actively adapt to the puncture process on the fuzzy boundary where puncture occurs.

Observe the puncture stages from 0.8 s to 0.9 s. At the stage when the adaptive smooth switching algorithm perceives that puncture occurs, the weight factor ρ rapidly rises to 1, and the weight factor ρ reaches the maximum value.

In addition, the puncture force rapidly decreases at about 0.3 s, that is, the stage when the cell membrane is penetrated. The weight factor ρ also rapidly decreases because of the short time of this stage, changing from 1 to 0 instantaneously.

To sum up, the proposed adaptive smooth switching algorithm has advantages. It can actively adapt to the fuzzy boundary of puncture occurrence, realize a rapid

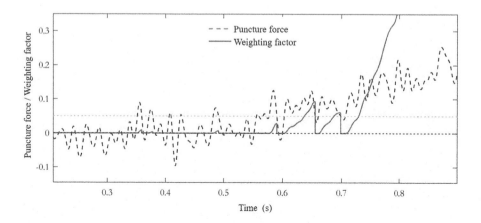

FIGURE 8.5   Partially enlarged diagram of weight factor ρ change simulation curve.

increase in the process of determining puncture occurrence, and also quickly fall back to 0 in the puncture completion stage. Therefore, the proposed adaptive smooth handover algorithm can effectively meet the requirements of the handover algorithm in the puncture process through computer simulation.

Furthermore, the weight factor ρ is a process variable, which represents the strength of the role of force and position controls when the force and position controls are in a coupled state.

8.4   OVERALL FLOW OF FORCE POSITION HYBRID CONTROL

The flow chart of force position hybrid control during cell puncture is shown in Figure 8.6. The laser displacement sensor and force sensor, respectively, collect displacement and force signals and provide feedback signals for the position controller and force controller, respectively. The key strategy in this flowchart is the adaptive smooth switching algorithm and the key actuator is the cell puncture mechanism. The adaptive smooth switching algorithm plays the role of "intermediate coordinator" in the action mechanism of the position controller and the force controller, as well as provides a unified excitation signal for the action of the cell puncture mechanism.

8.5   EXPERIMENTAL STUDY OF FORCE POSITION HYBRID CONTROL

Herein, computer simulation and hardware-in-the-loop simulation experiments [10] will be conducted for the proposed controller to investigate the performance of the proposed adaptive smooth switching algorithm and the force position hybrid controller.

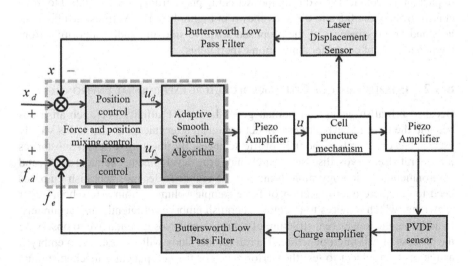

FIGURE 8.6   Flow chart of force position mixing control during cell puncture.

8.5.1 COMPUTER SIMULATION EXPERIMENT OF CELL MICROPUNCTURE

In this computer simulation experiment link, the software and hardware environments of the computer simulation experiment are the same as that in Chapters 3–6.

Given that the position tracking experiment and the micro-force tracking experiment have been comprehensively and deeply studied in Chapters 7 and 6, respectively, the computer simulation experiment will focus on the control effect of the adaptive smooth switching algorithm in the force position hybrid control.

Considering the effects of damping and stiffness, the contact force of cells in the computer simulation experiment is set as:

$$f_e = 0.3\dot{x} + 0.4\left(x - x_e\right) \tag{8.5}$$

where x_e is the distance from the glass microneedle to the cell membrane surface at the initial stage.

The expected force is set to:

$$f_d = 0.48t_{\text{puncture}} \tag{8.6}$$

where t_{puncture} is the time of the puncture stage.

The simulation process of the computer simulation experiment refers to the change strategy of the weight factor in the force level mixing control is shown in Figure 8.3; meanwhile, the force level mixing control flow chart in the cell micro puncture process is shown in Figure 8.6.

The relationship between the contact force and the expected force of the glass microneedle can be seen from the computer simulation experiment, as shown in Figure 8.7. Before the cell membrane is penetrated, the contact force always oscillates up and down around the expected force. Meanwhile, there is no additional impact on the contact force during the switching phase from position control to force control from about 0.5 s to 0.8 s, as shown in Figure 8.8. This verifies the effectiveness and progressiveness of the proposed adaptive smooth handoff algorithm from the perspective of computer simulation experiments.

8.5.2 ESTABLISHMENT OF CELL MICROPUNCTURE EXPERIMENTAL ENVIRONMENT

Because zebrafish genes and human genes have 87% affinity [11], zebrafish has become the third largest vertebra model organism after mice and rats and is widely used in basic research in the field of life sciences. By injecting different substances into zebrafish embryos, disease research, new drug development, chemical safety, and environmental toxicology monitoring can be realized. Because zebrafish embryos need to complete microinjection of large sample volume within a few hours after spawning, which requires to puncture zebrafish embryos efficiently and accurately. In addition, the transparent structure of zebrafish embryos is conducive to observing the process of cell micropuncture. Therefore, this study will take zebrafish embryos as the research object to test the performance of the cell puncture mechanism and robust controller.

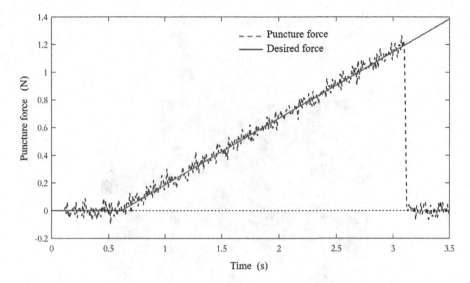

FIGURE 8.7   Contact force and expected force curve of glass microneedles in a computer simulation experiment.

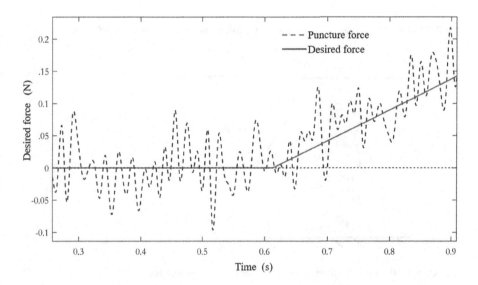

FIGURE 8.8   Partially enlarged view of contact force and expected force curve of glass microneedle at switching stage.

Build the experimental environment for micropuncture of zebrafish embryos, as shown in Figure 8.9. The cell puncture mechanism was installed on an inverted biological microscope (model ba1000, from Chongqing Optical Instrument Factory). Zebrafish embryos are located in the field of vision of the electron microscope (resolution: 300000 pixels, from Shenzhen Zongyuan Weiye Technology Co., Ltd.), and

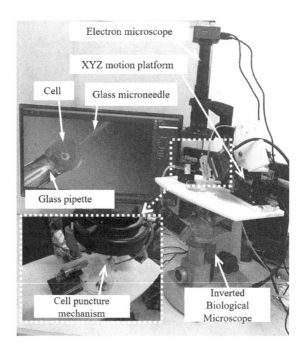

FIGURE 8.9   Experimental environment for micropuncture of zebrafish embryos.

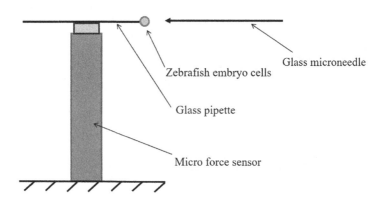

FIGURE 8.10   Layout schematic diagram of the micro-force sensor.

the puncture process is played in real time through the computer screen. Zebrafish embryos are adsorbed on the end of the glass straw by negative pressure, and the end of the glass straw is polished to avoid scratching the zebrafish embryos. To avoid the damage to zebrafish embryos caused by cell micropuncture, the end diameter of the glass microneedle (model: b150-86-7.5, from Chengdu chuanhuada scientific instrument factory) is 5 μm. It penetrates the cell membrane under the driving of the cell puncture mechanism, reaches a specific cell site, and then exits the cell membrane.

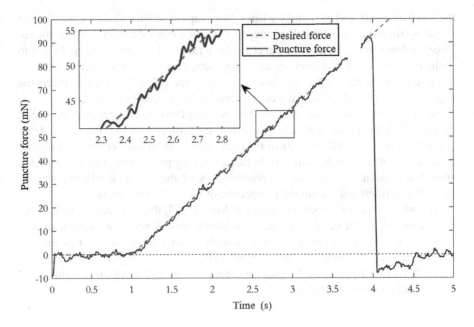

FIGURE 8.11   Change curve of puncture force during cell micropuncture.

The arrangement of the micro-force sensor is shown in Figure 8.11. The micro-force sensor is set in a vertical state to eliminate the influence of gravity on the sensor. The glass pipette is fixed on the micro-force sensor, and the glass microneedle punctures the cells in the direction perpendicular to the micro-force sensor to obtain the micro-force signal.

8.5.3   HARDWARE-IN-THE-LOOP SIMULATION EXPERIMENT OF CELL MICROPUNCTURE

The cell micropuncture experiment is completed using the adaptive smooth switching algorithm of force position hybrid control constructed by Eq. (8.1), combined with the position controller of Eq. (8.2) and the micro-force tracking controller of Eq. (8.3). As shown in Fig. 8.11, it is the change curve of puncture force in the cell micro puncture experiment. Combined with the results of computer simulation experiments, the physical and computer simulation experiments have been well verified.

At the beginning of the cell micro puncture experiment, the glass microneedle did not contact the cell surface; hence, the data collected by the resistance strain gauge micro-force sensor fluctuated at the zero point; as the glass microneedles slowly contact the cell membrane and extrude the cell membrane, the contact force gradually increases. In this process, with the ramp signal as the desired force, the puncture force can be accurately tracked by the micro-force tracking controller. Moreover, the sensor can accurately sense the time when the cell membrane is penetrated (around the fourth second), and the puncture force drops rapidly. Although the expected force signal is still rising, the controller enters the position control model from the micro-force tracking mode. During the puncture force rapidly falling back from 92 mN to 0, the body structure of the resistance strain gauge type micro-force sensor vibrates because of the inertia, hence, the data collected by the micro-force sensor enters the negative area, and subsequently, a certain amplitude of vibration occurs.

As shown in Figure 8.12, from the hardware-in-the-loop simulation experiment, the controller enters the switching phase from position control to force control from about 0.9 s to 1.1 s. Although there is always noise signal interference, the contact force does not change dramatically in this switching phase; however, it achieves a smooth transition. This proves the practicability of the proposed adaptive smooth handoff algorithm again from the perspective of physical experiments.

The whole puncture process is shown in Figure 8.13, the glass microneedles contact the outside of the cell membrane. Meanwhile, the cell puncture mechanism continuously increases the puncture force, and the glass microneedle causes the cell membrane to deform more greatly as shown in Figures 8.13 (b) and (c). Moreover, in Figure 8.13 (d), the head of the glass microneedle enters the zebrafish embryo and reaches the designated position.

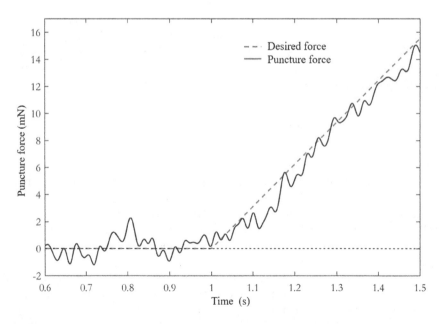

FIGURE 8.12 Partially enlarged view of contact force curve of glass microneedle at switching stage.

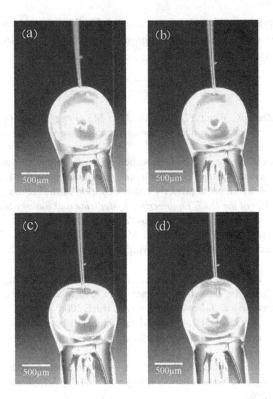

FIGURE 8.13 Photos of zebrafish embryos taken by electron microscope for micropuncture. (a) Glass microneedle contacted the cell membrane surface. (b) Extrusion of cell membrane with glass microneedle. (c) Glass microneedle further pressed the cell membrane surface. (d) Glass microneedle penetrated into the cell membrane.

8.6 CONCLUSION

The force position hybrid control strategy is introduced into cell micro puncture to effectively suppress the chattering and impact at the beginning of cell puncture and sense the instant state of successful puncture. The main achievements of this chapter are as follows:

An adaptive smooth switching algorithm is proposed in this chapter, which enables the controller to achieve smooth switching under different control modes. The adaptive smooth switching algorithm successfully couples force and position controls and brings the motion control of cell puncture mechanism based on fractional nonsingular terminal sliding mode and the micro-force tracking control of cell puncture into a unified control system. The automated switching between the force control subspace and the position control subspace is realized under the action of the excitation voltage signal, and the switching process is ensured to be smooth in transition without effect and sudden change.

Through the micropuncture experiment on zebrafish embryo cells, it is verified that the proposed adaptive smooth switching algorithm can realize the smooth switching from position control to force control. In addition, there is no chattering or impact in the process of gradual change of contact force, thus realizing accurate micro-force tracking control on cells.

REFERENCES

1. Zhao LQ, Zhao X, Wang YL, Sun MZ, Cui MS, Feng JZ, Lu GZ. Study on automatic toggle of oocytes in nuclear transplantation. *Control Engineering* 2012; 19(03): 389–393.
2. Liu LC, Lu GZ, Weng CH, et al. Research on micromanipulator and its microscopic visual servo control system. *High Technology Communication* 2001; 11(6):56–58.
3. Kim DH et al. Mechanical property characterization of the zebrafish embryo chorion. In: *Proceedings of the IEEE Engineering in Medicine and Biology Society, 26th Annual International Conference (IEMBS)* 2004; 2: 5061–5064.
4. Reed J, Ramakrishnan S, Schmit J, et al. Mechanical interferometry of nanoscale motion and local mechanical properties of living zebrafish embryos. *ACS Nano*, 2009; 3(8): 2090–2094.
5. Abdossalami A, Sirouspour S. Adaptive control of haptic interaction with impedance and admittance type virtual environments. In: *Symposium on Haptic Interfaces for Virtual Environment & Teleoperator Systems*. IEEE 2008: 145–152.
6. Zhong H, Li X, Gao L, et al. Toward safe human-robot interaction: a fast-response admittance control method for series elastic actuator. *IEEE Transactions on Automation Science and Engineering* 2021; 19(2): 919–932.
7. Poureetezadi SJ, Donahue EK, Wingert RA. A manual small molecule screen approaching high-throughput using zebrafish embryos. *Journal of Visualized Experiments JOVE* 2014; 93: e52063.
8. Huang H, Lindgren A, Wu X, et al. High-throughput screening for bioactive molecules using primary cell culture of transgenic zebrafish embryos. *Cell Reports* 2012; 2(3): 695–704.
9. Yu X, Dong S, Chong L. Penetration force measurement and control in robotic cell microinjection. *IEEE/RSJ International Conference on Intelligent Robots & Systems*. IEEE, 2009.
10. Haibin D, Daobo W, Xiufen Y. Realization of nonlinear PID with feed-forward controller for 3-DOF flight simulator and hardware-in-the-loop simulation. *Systems Engineering and Electronic Technology* 2008; 19(2): 4.
11. Amsterdam A, Sadler KC, Lai K, et al. Many ribosomal protein genes are cancer genes in zebrafish. *Plos Biology* 2004; 2(5): E139.

9 Automated Cell Biopsy Utilizing Micropuncture Technique*

9.1 INTRODUCTION

To understand normal physiology and functional mechanism of cells in response to disease or injury, the molecular biology of individual cells must be explored. Cellular functions, such as metabolism, cell motility, and gene expression, are greatly affected by the properties of intracellular structures and organelles. Technologies that are able to explore genetic factors and molecules involved in disease evolution at a single cell level become increasingly important for the investigation into the molecular basis of proliferation, cancer cell transformation, and metastasis. Studies on bulk tissues only provide a statistical average of several actions taken place in different cells. Single cell investigation may disclose that genetic changes caused by tumorigenesis-related signaling pathways lead to the change of a healthy cell to a cancerous cell [1]. Therefore, biological analysis at the single cell level provides information on individual cells and has thereby attracted worldwide attention in recent years. Single cell surgery at the subcellular level, such as the manipulation or removal of subcellular components or/and organelles [2] from a single cell, is increasingly needed to study diseases and their causes [3]. The techniques associated with single cell surgery include preimplantation and diagnosis [4], understanding the organelle and subcellular activities [1], cloning [5], to name a few. Highly precise micro-manipulation techniques that are able of manipulating and interrogating cell organelles and components must be developed to benefit the rapid development of new cell-based medical therapies, hence facilitating an in-depth understanding of cell dynamics, cell component functions, and disease mechanisms.

Several micro/nano-manipulation tools for intracellular-level surgery have been developed over the past few decades. For instance, a scanning-ion-conductance-microscopy-based nano-biopsy system was developed to extract femtoliter samples of intracellular content to analyze mRNA and mitochondrial DNA. This technique utilized electro-wetting to uptake samples of cytoplasm into a glass nanopipette in a minimally invasive way, and then used high-throughput sequencing technology to analyze the extracted cellular material [6]. AFM tips were utilized to extract specific mRNA from live cells [7] and nanorobots were fabricated for puncturing cells [8]. Similarly, carbon nanotubes placed at the tip of a glass pipette were used at the single cell level for transferring fluids, interrogating cells, and performing optical and electrochemical diagnostics [9].

* Chapter by Dr. Adnan Shakoor, Control and Instrumentation Department, King Fahd University of Petroleum & Minerals, Dhahran, Saudi Arabia.

DOI: 10.1201/9781003294030-9

These subcellular-level cell surgery tasks are currently performed by highly skilled operators. Moreover, operator fatigue leads to reduced accuracy, repeatability, and high error rates. To gain significant practical experience in these complex cell surgery operations, operators must receive intensive and lengthy training. The automation of these complex single cell surgery tasks is in high demand to reduce the probability of contamination resulting from human errors, labor-intensive work, process uncertainty, and variable outcomes. To manipulate biological cells, several robotic micromanipulation systems have been reported in literatures [10–21]. For instance, embryo biopsy, an important step prior to preimplantation genetic diagnosis, is a complex single cell surgery task in which the zona pellucida is dissected with a glass micropipette, followed by blastomere biopsy from an embryo. Several studies have examined the automation of this process, which include force measurements during micropipette insertion [22], contact detection [23], automatic position selection for zona pellucida dissection [24], and automated micropipette control and embryo rotation tracking [25], to name a few. Similarly, several studies have attempted to automate this mechanical enucleation process by using different micromanipulation tools, such as micropipettes [26], magnetically driven microtools [27], and microfluidic chips [28]. Note that all the aforementioned approaches were applied to large-scale cells of ~80 μm. Single cell biopsy for small cells, such as many human cells (~20 μm), is more challenging because of their small size, irregular shape, and flexible membranes, all of which demand high-precision manipulation.

Cell puncture is a process of inserting foreign materials into cells. Several cell puncture systems that allow semi [29–35] and fully automatic [23, 36–39] injection processes have been proposed. However, cell biopsy differs from cell injection in several aspects. First, cell puncture is accomplished at an arbitrary location within the cell, usually at the center [36, 39]. In cell biopsy, the micropipette tip must be precisely inserted inside the cell at the position of the desired organelle. Precisely accessing the spatial position of the organelle within the cell, especially for small suspended cells such as most of human cells in dimensions of ~20 μm, is challenging. Thus, a highly precise positioning control is needed for cell biopsy. Second, cell biopsy requires a number of extra operations, such as organelle aspiration control inside the micropipette and cell holding control during extraction. Therefore, cell organelle biopsy is more complex than cell injection.

In this chapter, an automated organelle biopsy system for small dimension cells was proposed after cell puncture technique. The main contributions of this chapter are described as follows. First, a strategy for single cell organelle is developed for small cells with dimensions of less than 20 μm. Second, to meet high requirement for organelle positioning accuracy, a microfluidic single cell patterning device is designed, which can compress and pattern cells inside rectangular channels in a 1-D array, hence simplifying the automated process. Third, under a visual-based robust control algorithm, the approach can be applied to biopsy a broad range of cells with high manipulation accuracy while addressing model uncertainties and external disturbances. Experiments of automated biopsy of mitochondria and nucleus from human acute promyelocytic leukemia cells (NB4) and human dermal fibroblast (HDF) cells are performed to demonstrate the effectiveness of the proposed strategy. Extraction rate of mitochondria biopsy and cell viability for different size of micropipette is analyzed. The functionality of mitochondria and the viability of the biopsied cell after removal of mitochondria are also investigated.

The rest of the chapter is organized as follows: Section 9.2 introduces the microfluidic cell patterning device; Section 9.3 describes the organelle biopsy process and control strategy; Section 9.4 demonstrates the effectiveness of the automatic organelle biopsy approach through experimental studies; and finally, conclusions of this chapter are presented in Section 9.5.

9.2 MICROFLUIDIC CELL PATTERNING STRATEGY

9.2.1 CELL PATTERNING

To simplify the cell manipulation process, cells are patterned in a 1D array. PDMS-based microfluidic cell patterning device (MCPD) is designed by placing many rectangular shaped channels in parallel. This design eliminates the need for a second manipulator (for cell holding) and simplifies the cell/organelle searching and extraction processes. Figure 9.1 presents the schematic design of the microfluidic device. MCPD consists of an aspirating layer and a cell patterning layer. The arrangement of these two layers produces a rectangular hole to pattern and hold each cell firmly. A small cylindrical gap called the "store" is added near the front of the channel to hold and store the biopsied sample as shown in Figure 9.1. Two nearby channels are separated by a distance of 100 μm to ensure that only one channel and a "store" can

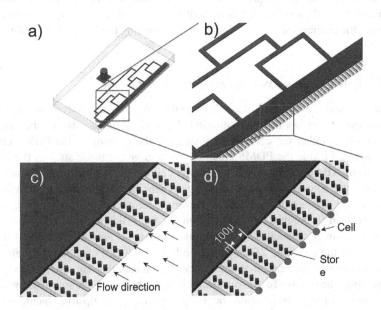

FIGURE 9.1 Schematic design of the cell patterning microfluidic device. a) Full view of the microfluidic device. b) Close-up view of the inset. c) Flow direction. d) Cell trapping. (Source: A. Shakoor, M. Xie, T. Luo et al. / IEEE Transactions on Biomedical Engineering 66 (2019) 2210–2222, with permission.)

be visualized within the field of view, thereby simplifying the image processing for detecting the position of the organelle for a cell present in the field of view.

The microfluidic device was fabricated with soft lithography technology. First, a 5 μm layer of SU-8 photoresist (GM1050, Gersteltec Sarl) was spin coated on a silicon wafer and prebaked on a hotplate (AccuPlate, Labnet). The structure was exposed to UV light with the mask on top. The second layer of 15 μm thickness was fabricated following the same process as above. Then a mixture of PDMS and curing agent with a ratio of 10:1 was dropped onto the master mold and baked in an oven for 1.5 hours, and finally peeled off the master mold.

After fabrication, the microfluidic cell patterning device was plasma bonded with a glass slide, was rinsed with a cell culture medium, and exposed to UV light for half an hour. Without this step, the liquid may not flow smoothly through the channels and unwanted bubbles may be generated in front of the channels. The cell patterning device was attached to a glass slide holder, which was fixed on the X–Y–Z positioning stage. To generate a fluid flow, the outlet of the cell holder was attached to a digital micro-injector through polyethylene tubing. A pressure of −124.6 Pa was initially applied, and the cells were placed near the channels with a pipette. Within a few minutes, the cells were trapped in front of the channels and the pressure was adjusted to either squeeze the cell inside the channel or maintain its position at the channel opening. When cells were present at the channel opening, a pressure of 24.9 Pa was considered sufficient to maintain the position of the cell. Increasing the pressure to 250 Pa for approximately 10 seconds would aspirate 90% part of the cell into the channel, while reducing the pressure to 24.9 Pa would maintain the cells in this state. To ensure that only a single cell is present within the field of view, the non-trapped cells near the channels were removed by gently flushing the cell culture medium with a pipette.

9.2.2 CELL COMPRESSION FOR ORGANELLE POSITIONING

Figures 9.2 and 9.3 illustrate the cell assembly by using hydrodynamic drag force inside channels for organelle positioning. Depending on the size of the organelle to be biopsied, either the cell can be trapped at the opening of the PDMS channel or compressed inside the PDMS channel. For a large organelle such as the nucleus (7–9 μm) of NB4 cells, its position within a cell of ~20 μm can be easily accessed. Therefore, the automated biopsy of a large organelle can be achieved after holding the cell at the opening of the channel. However, for a small organelle, the cell must be compressed to confine the 3D position for the following reasons. First, unlike the nucleus, small organelles such as mitochondria (~2 μm) are distributed in a 3D space within the cell. The 2D images [see Figure 9.3 (a)] obtained through florescence microscopy are combinations of stacked images that are obtained at distinct depths along the z-axis. To perform mitochondria biopsy in an automatic way, it is challenging to precisely position the 1–2μm micropipette tip to coincide with the position of the mitochondria along the z-axis [see Figure 9.3(b)], which requires the use of complex image processing algorithms to obtain the accurate position of small organelles along the z-axis. Second, compressing the cell helps to accurately determine the negative pressure, applied by the motor-controlled injector, needed to

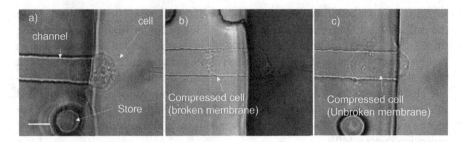

FIGURE 9.2   Cell (NB4) patterning by a microfluidic device. (a) The cell is trapped in front of the channels. (b) The cell is compressed in a channel of 2 μm height. Arrow head pointing the bubble formation due to membrane opening. (c) Compressed cell, with unbroken membrane, in a channel with height of 5 μm. Scale bar is 15 μm. (Source: A. Shakoor, M. Xie, T. Luo et al. / IEEE Transactions on Biomedical Engineering 66 (2019) 2210–2222, with permission.)

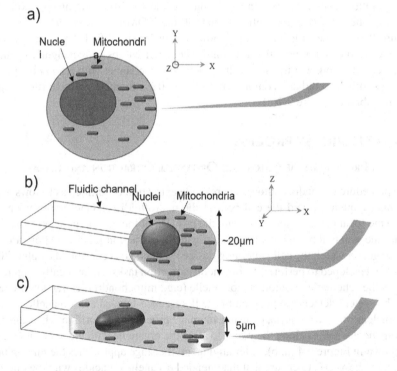

FIGURE 9.3   Confining the position of the mitochondria along the z-axis by compressing the cell. (a) 2D camera image view (top view). (b) 3D representation of the cell before compression. (c) 3D representation of the cell after compression inside the fluidic channel. (Source: A. Shakoor, M. Xie, T. Luo et al. / IEEE Transactions on Biomedical Engineering 66 (2019) 2210–2222, with permission.)

aspirate the mitochondria into the micropipette. If the micropipette tip is inserted into the cell where the mitochondrion is positioned relatively far from the micropipette tip along the z-axis (which is common for cells in their suspended state), then a large volume of cytoplasm must also be aspirated along with the mitochondria, which can cause significant damage to cell. Without restricting the position of the small organelle along the z-axis, the negative pressure applied through the computer-controlled injector to aspirate the organelle may vary for each cell, thereby making the biopsy results inconsistent or a human operator would be required to control the amount of aspiration content [40].

Considering the adherent cells that contain organelles in a confined position along the height of the cell, a careful compression of suspended cells can restrict the position of small organelles without damaging cell. Therefore, by compressing the cell, the distance between the micropipette tip and the small organelle along the z-axis is restricted by the height of the channel, which is 5 μm only, as shown in Figure 9.3(c). Furthermore, the beveled micropipette along with the compressed mode of the cell, which tightens the cell membrane, facilitates the perforation of the cell membrane. For the above-mentioned reasons, instead of using complex image processing algorithms to determine the 3D position of the mitochondria as well as complex control of the motion along the z-axis and of the biopsy injector, by aspirating approximately 95% of the cell into the soft channel of the PDMS microfluidic device [see Figure 9.3(c)], the organelle biopsy process can be simplified. Experiments were conducted to observe how the different heights and widths of rectangular channels affect cell trapping and its vitality. It is found that a channel with a 5 μm height and 15 μm width can hold and compress the cell of 20 μm diameter without rupturing its cell membrane [Figure 9.2(c)].

9.3   CELL BIOPSY PROCESS

9.3.1   Procedures of Automatic Organelle Extraction and Release

The procedure to conduct automatic organelle extraction is outlined as follows. First, the micropipette tip and the cell held in the microfluidic channel are brought into the same focal plane. After alignment, the position of the micropipette remains fixed along the z-axis throughout the whole organelle extraction process. Afterward, the x-y stage, on which the cell patterning device is attached, is controlled by algorithms that are developed to perform the organelle extraction tasks automatically. Figure 9.4 shows the schematic of automatic organelle (e.g., mitochondria) extraction process.

The system detects the position of the florescence-labeled cell organelle, such as the nucleus or mitochondria, by using an image processing algorithm. The stage then automatically moves along the y-axis to align the organelle with the micropipette tip, as shown in Figure 9.4 (a, b). After alignment, the stage approaches the micropipette tip along the x-axis to ensure that the intended organelle coincides with the micropipette tip [Figure 9.4(c)]. Then, the microinjector aspirates the organelle into the micropipette tip [Figure 9.4(d)].

The controller then moves the stage away from the micropipette to extract the micropipette from the cell [Figure 9.4(e)]. The aspirated organelle can be placed in

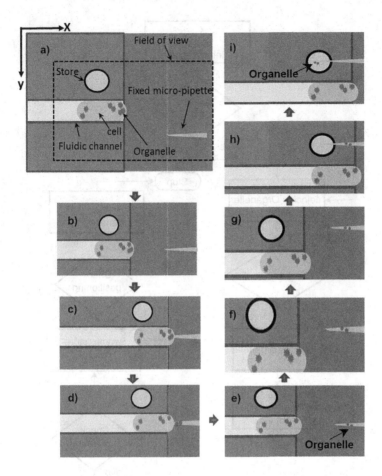

FIGURE 9.4 Schematic of organelle extraction. (a–e) Steps of the organelle extraction process and (f-i) steps of releasing the organelle into the store. (Source: A. Shakoor, M. Xie, T. Luo et al. / IEEE Transactions on Biomedical Engineering 66 (2019) 2210–2222, with permission.)

the "store" [see Figure 9.4(f–i)] automatically for further analysis. To release the organelle inside the "store," the stage moves 15 μm along the y-axis to bring the "store" within the microscope field of view [Figure 9.4(f)]. The center of the store is detected with an image processing technique, and the stage moves the cell patterning device along the y-axis to align the micropipette with the center of the store [Figure 9.4(g)]. The stage further moves along the x-axis to perforate the micropipette tip into PDMS such that the center of the store coincides with the micropipette tip [Figure 9.4(h)]. Finally, the microinjector ejects the organelle into the store, automatically [Figure 9.4(i)]. Figure 9.5 presents the flow chart for this process.

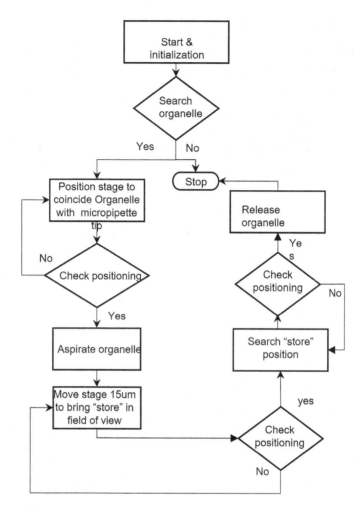

FIGURE 9.5 Flow chart of the automated biopsy process. (Source: A. Shakoor, M. Xie, T. Luo et al. / IEEE Transactions on Biomedical Engineering 66 (2019) 2210–2222, with permission.)

9.3.2 Motion Control

The dynamic equation of a 3-DOF motorized stage (see Figure 9.6) can be determined by using the following Lagrange's equation of motion:

$$M\ddot{q} + B\dot{q} + G(q) = \tau \tag{9.1}$$

where $M \in \Re^{3 \times 3}$ denotes the inertial matrix of the system, $q = [q_x, q_y, q_z]^T \in \Re^{3 \times 1}$ is the stage position coordinate, $B \in \Re^{3 \times 3}$ denotes the damping friction coefficient, $G = [0, 0, mg] \in \Re^{3 \times 1}$ models the gravity force, and $\tau = [\tau_x, \tau_y, \tau_z]^T$ is the input torque. Both M and B are diagonal and positive definite.

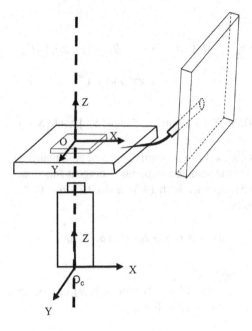

FIGURE 9.6 Coordinate frames of the X–Y–Z stage and camera. (Source: A. Shakoor, M. Xie, T. Luo et al. / IEEE Transactions on Biomedical Engineering 66 (2019) 2210–2222, with permission.)

Given the model uncertainty and external disturbances (e.g., mechanical vibration), the dynamic model of the motorized stage can be generalized as follows:

$$\left(M_0 + \Delta M\right)\ddot{q} + \left(B_0 + \Delta B\right)\dot{q} + \left(G_0 + \Delta G\right) + \tau_d = \tau \tag{9.2}$$

where M_0, B_0, and G_0 denote the nominal values of the system parameters, ΔM, ΔB, and ΔG represent the system uncertain values, and τ_d denotes the external disturbance torque.

A sliding variable is defined as follows:

$$S_r = \dot{q} - \dot{q}_r = \dot{q} - \left(\dot{q}_d - \alpha q_e\right) = \dot{q}_e + \alpha q_e \tag{9.3}$$

where $q_e = q - q_d$ is the position error, and α is the positive constant.

The dynamic model of the motorized stage in Eq. (9.1) can be expressed in terms of S_r as follows:

$$M\dot{S}_r + BS_r + G(q) = \tau + M\left(-\ddot{q}_d + \alpha\dot{q}_e\right) + B\left(-\dot{q}_d + \alpha q_e\right) \tag{9.4}$$

Then, a robust sliding controller is designed as follows to achieve trajectory tracking:

$$\tau = \tau_0 + \tau_s \tag{9.5}$$

where

$$\tau_0 = -M_0\left(-\ddot{q}_d + \alpha\dot{q}_e\right) - B_0\left(-\dot{q}_d + \alpha q_e\right) + G_0 \tag{9.6}$$

$$\tau_s = -KTanh\left(S_r\right) \tag{9.7}$$

$$Tanh\left(S_r\right) = \left[\tanh\left(S_{r1}\right), \tanh\left(S_{r2}\right), \tanh\left(S_{r3}\right)\right]^T \tag{9.8}$$

where $K = diag\{k_1, k_2, k_3\}$ is the positive control gain matrix, $S_r = [S_{r1}, S_{r2}, (S_{r3})]^T \in R^{3 \times 1}$, and $Tanh(\cdot)$ is the standard hyperbolic tangent function.

Substituting (9.5) together with (9.6) and (9.7) into (9.2) yields the following closed-loop dynamics:

$$M_0\dot{S}_r + B_0S_r + KTanh\left(S_r\right) + \Delta f = 0 \tag{9.9}$$

where $\Delta f = \Delta M\ddot{q} + \Delta B\dot{q} + \Delta G + \tau_d$.

The stability of the closed-loop dynamics (9.9) can be analyzed by defining a Lyapunov function candidate as follows:

$$V = \frac{1}{2}S_r^T M S_r \tag{9.10}$$

By differentiating V with respect to time and utilizing the closed-loop equation (9.9), we have

$$\begin{aligned}\dot{V} &= S_r^T M \dot{S}_r \\ &= -S_r^T K S_r - S_r^T\left(KTanh\left(S_r\right) + \Delta f\right) \\ &= -S_r^T K S_r - \sum_{i=1}^{3} S_{ri}\left[k_i Tanh\left(S_{ri}\right) + \Delta f_i\right]\end{aligned} \tag{9.11}$$

When $S_{ri} > 0$, given $k_i > |\Delta f_i|$ for $i = 1, 2, 3$, we have

$$\sum_{i=1}^{3} S_{ri}\left[k_i Tanh\left(S_{ri}\right) + \Delta f_i\right] > 0 \tag{9.12}$$

Note that k_i is easily chosen to be larger than $|\Delta f_i|$, based on the fact that $\dot{q}$ and $\ddot{q}$ are not so large in cell manipulation process and hence Δf_i is bounded.

When $S_{ri} < 0$, given $k_i > |\Delta f_i|$, we have

$$\sum_{i=1}^{3} S_{ri}\left[k_i Tanh\left(S_{ri}\right) + \Delta f_i\right] > 0 \tag{9.13}$$

Therefore,

$$\dot{V} \leq 0 \tag{9.14}$$

By using the LaSalle invariance principle, we conclude that the system states are eventually driven into the sliding surface $S_r = 0$, and thus the tracking errors q_e and $\dot{q}_e$ converge to zero.

9.3.3 BIOLOGICAL TESTS

Biological tests are further conducted on the extracted organelles to characterize their biological functionality. The mitochondria membrane potential has been widely used as an indicator of mitochondria heath [41–43]. A significant reduction in membrane potential of mitochondria is an indicator of mitochondria death. By using the proposed organelle biopsy system, it must be verified that the membrane potential of the extracted mitochondria can remain the same as that before the mitochondria biopsy therefore demonstrating that the extracted mitochondria has survived. The invasiveness of biopsy process on cell by calcium imaging before and after the mitochondria biopsy was reported in [6]. We can demonstrate that membrane potential of the extracted mitochondria and calcium flux of the remained cell after mitochondria biopsy remained almost unchanged, which are reported in detail in Section 9.4.

9.4   EXPERIMENTS

9.4.1 MATERIAL PREPARATION

NB4 and HDF cells were used in the organelle biopsy experiments. NB4 cells were maintained in a petri dish containing RPMI 1640 with 2 mM L-glutamine inside a humidified incubator with an atmosphere of 37 °C and 5% CO_2. HDF cells were kept in Dulbecco's modified Eagle's medium (DMEM, Gibco), 100 U/mL penicillin supplemented with 10% fetal bovine serum (FBS, Gibco), and 100 U/mL streptomycin inside a humidified incubator with an atmosphere of 37 °C and 5% CO2. The cells were then stained with MitoTracker® Red CMXRos (M7512, Thermo Fisher Scientific) and Hoechst (33342, Thermo Fisher Scientific) for the florescence labeling of mitochondria and nucleus, respectively. The HDF cells were treated with trypsin (Sigma) for detachment from the petri dish, while the NB4 cells were directly used in the suspended state.

A glass micropipette (O.D. 1 mm and I.D. 0.78 mm) was pulled using a micropipette puller (P-2000, Sutter Instrument), and a gradual taper with a final tip size of 0.5 μm was obtained. To ensure the perforation of the flexible cell membrane and minimize damage to the cell (for the mitochondria biopsy), the micropipette tip was beveled using a BV-10 microelectrode beveller (Sutter Instrument) at 30° with a final tip size of about 1–3um for mitochondria biopsy and 5–8 um for nucleus biopsy. The micropipette was bent 4–5 mm from the tip to ensure that the tip can access the cell when the latter is docked in the cell patterning device.

9.4.2   DETECTION OF MITOCHONDRIA, NUCLEUS AND STORING SPACE

To detect the position of the mitochondria region, an image was captured through a combination of florescence and bright field microscopy as shown in Figure 9.7(a) before being converted into an HSV plan [Figure 9.7(c)]. Changing the HSV values created a corresponding threshold image as shown in Figs. 9.7(c), and these values remained the same for a fixed brightness and exposure time. The mitochondria region, hue, saturation,

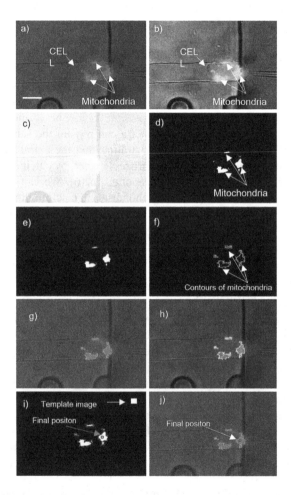

FIGURE 9.7   Automatic mitochondria detection. (a) Cell compressed inside the channel and stained with a Mito tracker red as indicated by arrows. (b) Grayscale image of (a) (for better visualization). (c) Image in the HSV plane. (d) Image after applying thresholding. (e) Image after applying Gaussian and Median filters. (f–h) Extracting the contours from the thresholding image and imposing contours on the original image. (h) Grayscale representation of image (g). (i, j) Final selected position of mitochondria, as indicated by an arrow, after applying the template matching algorithm and using the template image shown in the inset of (i). Scale bar is 15 μm. (Source: A. Shakoor, M. Xie, T. Luo et al. / IEEE Transactions on Biomedical Engineering 66 (2019) 2210–2222, with permission.)

and value of the image were represented by HSV values of 0–10, 160–179, 0–255, and 0–179, respectively. By applying these HSV values on the HSV plan image, a thresholding image was obtained to differentiate the mitochondria, which is marked by red florescence, from the other parts of the image as seen in Figure 9.7(d). For verification purposes, the contours were extracted and imposed on the original image to confirm the accurate thresholding of the mitochondria region [Figure 9.7(f–h)]. To reduce the high frequency and pepper noise in the image, Gaussian and Median filters were applied [Figure 9.7(e)]. A small image with few white pixels as shown in the right corner of Figure 9.7(i) was utilized as a template image to detect the random position of the mitochondria's thresholding region. The normalized cross-correlation template matching method (CCORR_NORMED) was applied to determine the final detected position. The final position obtained via CCORR_NORMED is denoted by a small black square and arrowhead in Figure 9.7(g, h).

To detect the nucleus, the cell image was captured through a combination of florescence and bright field microscopy [Figure 9.8(a)] and was then converted into an

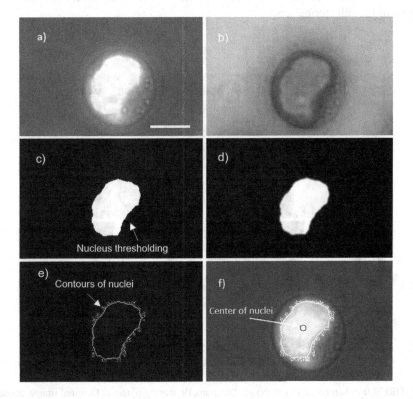

FIGURE 9.8 Automatic detection of the nucleus center. (a) Image of the cell stained with Hoechst 3342. The blue bright color represents the nucleus of the cell. (b) Threshold image. (c) Image after applying Gaussian low pass filter. (d, e) Contours extracted from the threshold image. (f) Detection of the center of the nucleus as represented by a small black circle (also pointed by an arrow) and imposing the contour on the original image. (Source: A. Shakoor, M. Xie, T. Luo et al. / IEEE Transactions on Biomedical Engineering 66 (2019) 2210–2222, with permission.)

HSV plan [Figure 9.8(b)]. With HSV values of 0–100 for hue, 0–255 for saturation, and 0–179 for value, a thresholding image was obtained to differentiate the nucleus, which is marked by blue florescence, from the rest of the image [Figure 9.8(c)]. The image was then passed through Gaussian and Median filters to reduce noise [Figure 9.8(d)]. To find the center of the nucleus, the nucleus contour was detected [see Figure 9.8(e)] and the center of the nucleus was then located by finding the central moment of the bounded area [Figure 9.8(f)].

A template matching algorithm was utilized to detect the organelle store space automatically, as shown in Figure 9.9. The image containing the cell storing space [Figure 9.9(a)] was converted into a grayscale image [Figure 9.9(b)]. The image then passed through a low-pass Gaussian filter, and thresholding was applied to convert the image into a binary image [Figure 9.9(c)]. A template image, as shown in the inset of Figure 9.9(a), was used to match the area in the binary image. CCORR_ NORMED was applied to locate the area with the highest matching probability. The rectangle in Figure 9.9(d) presents the results of CCORR_NORMED. The length of the vertices of this rectangle was used to determine the center of the rectangle or the center of the cell storing space.

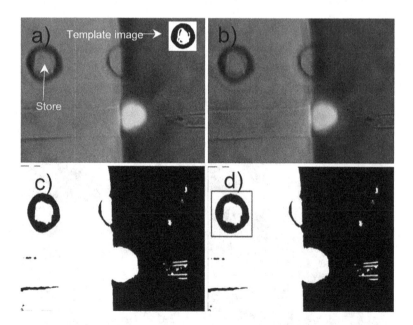

FIGURE 9.9 Automatic detection of the organelle storing place. a) Original image containing the store as indicated by an arrow. The inset template image is used to find the positions of the stores in the images. b) The image is converted from RGB to grayscale and filtered with a low-pass Gaussian filter to reduce noise. c) Threshold image of the grayscale image. d) Detection of the position of the organelle store by using a template matching algorithm as indicated by the rectangle around the store. (Source: A. Shakoor, M. Xie, T. Luo et al. / IEEE Transactions on Biomedical Engineering 66 (2019) 2210–2222, with permission.)

9.4.3 ORGANELLE EXTRACTION

To biopsy mitochondria and nucleus of single cells, successful extraction of these organelles from cells is the first important action. Figure 9.10 shows images of the automatic extraction of mitochondria from HDF cells. These images were converted into grayscale for better visualization. To perform mitochondria extraction in a minimally invasive way, a beveled and nearly cylindrical micropipette with a tip size of about 2–3 μm was utilized [Figure 9.10(a)]. The mitochondria position [Figure 9.10(b)] was automatically identified using the proposed image processing algorithm and was aligned with the micropipette tip along the x-axis, as shown in Figure 9.10(c). The cell patterning device then moved toward the micropipette tip until the micropipette tip coincided with the detected position of the mitochondria after perforating the cell membrane [Figure 9.10(d)]. The injector aspirated the mitochondria

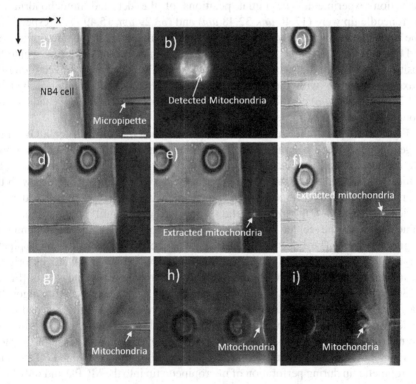

FIGURE 9.10 Automatic extraction of the mitochondria. Image is presented in grayscale for better visualization. (a) Micropipette and cell before biopsy. Scale bar is 10 μm. (b) Mitochondria detection. The small square pointed with an arrowhead shows the detected position of the mitochondria. (c) Moving the stage to the aligned micropipette with the detected mitochondria. (d) Moving the stage toward the micropipette. (e) Aspirated mitochondria. (f) Moving the stage back to extract the micropipette out of the cell and moving the stage 15 μm along the y-axis to bring the "store" within the field of view. (g) Moving the stage along the y-axis to align the micropipette with the center of the "store." (h) The micropipette reaching the center of the store. (i) Releasing the mitochondria. (Source: A. Shakoor, M. Xie, T. Luo et al. / IEEE Transactions on Biomedical Engineering 66 (2019) 2210–2222, with permission.)

into the micropipette [Figure 9.10(e)], and the cell patterning device moved toward its original position (along the x-axis) to extract the micropipette from the cell. To ensure that the "store" is completely present within the field of view, the stage moved 15 μm along the y-axis [Figure 9.10(f)] and was positioned along the y-axis to align the center of the "store" with the micropipette tip. The stage moved along the x-axis to coincide the center of the store with the micropipette tip. The injector ejected the aspirated mitochondria [Figure 9.10(f)] by applying positive pressure at the back of the micropipette.

Extraction of the mitochondria from NB4 cells was also conducted by following the same sequence as explained above, with the results presented in Figure 9.11. The aspirated mitochondria can be released outside the cell patterning device as shown in Figure 9.11(f).

Figure 9.12(a, b) shows the position error obtained during the mitochondria extraction experiment. The initial positions of the detected mitochondria and microneedle tip were (17.40 μm, 32.48 μm) and (33.29 μm, 15.49 μm), respectively. The stage moved along the y-axis to align the mitochondria with the micropipette tip and reached the (17.40 μm, 15.48 μm) position within 5 seconds, during which the position error along the y-axis converged to zero. The stage then moved along the x-axis to coincide with the micropipette tip. The stage eventually reached the (33.27 μm, 15.48 μm) position with a final position accuracy of 0.02 μm and 0.01 μm along the x and y axes, respectively.

Figure 9.13 shows the automatic extraction of nucleus from NB4 cell. Given the large organelle to cell size ratio, cell compression is not necessary for nucleus biopsy and the cell can be trapped at the channel inlet. The center of the nucleus was automatically identified using the image processing algorithm and was aligned with the micropipette tip along the x-axis, as shown in Figure 9.13(b). The cell patterning device moved toward the micropipette tip for perforation until the micropipette tip coincided with the center of the nucleus after perforating the cell membrane [Figure 9.13(c)]. The injector aspirated part of the nucleus into the micropipette [Figure 9.13(d)]. Afterward, the cell patterning device moved toward its original position along the x-axis, while the stage moved 15 μm along the y-axis [Figure 9.13(e)] to bring the "store" within the field of view. An image processing algorithm was used to detect the position of the organelle store space, and the cell patterning device was controlled to move along the y-axis to align the center of the store with the micropipette tip [Figure 9.13(f)]. Stage then moved along x-axis until the store center met the micropipette tip [Figure 9.13(g)]. Cell debris was cleared off from the micropipette tip during perforation of micropipette tip into the MCPD and only biopsied nucleus remained into the micropipette tip [Figure 9.13(g)]. The injector ejected the biopsied nucleus from the tip into the organelle store space [Figure 9.13(h, i)] by applying positive pressure.

The micropipette can be easily pulled out from the cell if the extracted organelle is small (e.g., mitochondria) but this task is challenging if the extracted organelle is large (e.g., nucleus). For aspirating large organelles, a greater negative pressure is applied to the micropipette during organelle aspiration, which may restrict the separation of the micropipette and the cell. For example, the nucleus, when occupying more than 60% of the cell volume, requires a relatively high negative pressure to be

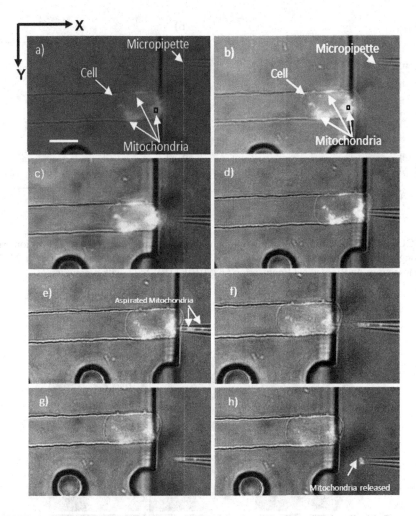

FIGURE 9.11 Automated extraction of the mitochondria from NB4 cell. (a) Detected position of the mitochondria shown with small rectangle and pointed with arrowhead. (b) Grayscale image of (a) for better visualization. (c) Micropipette aligned with the selected position of the mitochondria. (d) Moving the stage toward the micropipette. (e) Aspirated mitochondria. (f) Moving the stage back toward its original position. (g) Moving the stage 15 μm along the y-axis. (h) Releasing the mitochondria. Scale bar is 15 μm. (Source: A. Shakoor, M. Xie, T. Luo et al. / IEEE Transactions on Biomedical Engineering 66 (2019) 2210–2222, with permission.)

aspirated into the micropipette tip with a size ranging from 5 μm to 8 μm. If same amount of negative pressure is applied at the back of the fluidic channel, the cell will rupture during the extraction of the micropipette while some of its parts (cell debris) remain attached to the micropipette. Given the large size of the micropipette for nucleus biopsy, the biopsy of the nucleus is very invasive and the remaining part of the cell is considered cell debris. Consequently, instead of controlling the pressure at

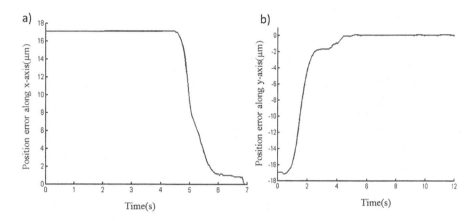

FIGURE 9.12 Position errors of the mitochondria extraction experiment. (a) Position error along x axis. (b) Position error along y axis. (Source: A. Shakoor, M. Xie, T. Luo et al. / IEEE Transactions on Biomedical Engineering 66 (2019) 2210–2222, with permission.)

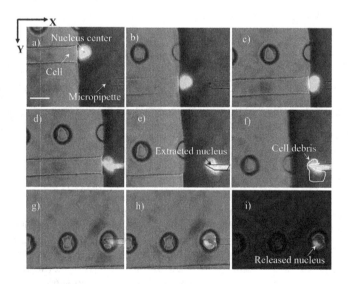

FIGURE 9.13 Automated extraction of nucleus from NB4 cell. (a) Automatic detection of the nucleus center via image processing. Scale bar is 15 μm. (b) The nucleus aligned with the tip of the micropipette by moving the stage along the y-axis. (c) Moving the stage toward the micropipette (along the x-axis). (d) Aspirating part of the nucleus. (e) Moving the stage toward its original position (along the x-axis) and 15 μm along the y-axis to bring the "store" within the field of view. (f) Moving the stage along the y-axis to align the micropipette with the center of the store. (g) Penetrating the micropipette tip by moving the stage along the x-axis. (h–i) Releasing the nucleus. (Source: A. Shakoor, M. Xie, T. Luo et al. / IEEE Transactions on Biomedical Engineering 66 (2019) 2210–2222, with permission.)

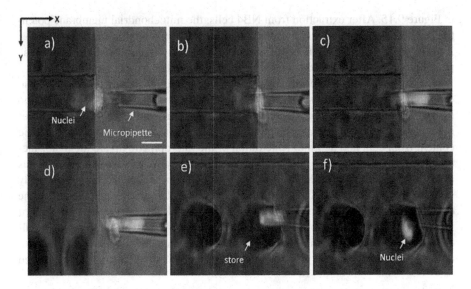

FIGURE 9.14 Automated extraction of the nuclei from HDF cell. (a) Aligning the nucleus with the tip of the micropipette by moving the stage along the y-axis. Scale bar is 15 μm. (b) Moving the stage toward the micropipette (along the x-axis). (c) Aspirating nucleus. (d) Moving the stage toward its original position (along the x-axis) and 15 μm along the y-axis to bring the "store" within the field of view. (e) Penetrating the PDMS by moving the stage along the x-axis. (f) Releasing the nucleus inside the store. (Source: A. Shakoor, M. Xie, T. Luo et al. / IEEE Transactions on Biomedical Engineering 66 (2019) 2210–2222, with permission.)

the back of the channel or extracting the micropipette from the cell, the cell debris is removed by penetrating the micropipette into PDMS. This rinsing process can be achieved automatically during the movement of the micropipette toward the store [see Figure 9.13(f–g)]. The negative pressure required at the back of the micropipette during organelle aspiration depends on the amount of the cellular content to be biopsied and the tip size of the micropipette. These control parameters were fixed through calibration experiments.

Figure 9.14 shows the experimental results of the nuclei extraction performed on the HDF cell in the same sequence as explained above for the nuclei biopsy of NB4 cell.

9.4.4 BIOLOGICAL TESTS ON EXTRACTED ORGANELLES AND THE REMAINED CELLS

Upon completion of organelle extraction, a number of biological tests on the extracted organelles and the remained cells were conducted. The functionality test of the extracted mitochondria was performed by observing the mitochondria membrane potential before and after the mitochondria biopsy. Cells were stained with JC-1 Dye, Mitochondrial Membrane Potential Probe (T3168, Thermo Fisher Scientific), prior to the biopsy experiment. According to the report by [41–43], depolarized mitochondria shows a change in the florescence emission from yellowish orange or reddish orange (depending on the filter type) to green. The testing results of our study were illustrated in Figure 9.15. After extraction from NB4 cells, the mitochondrial membrane potential remained unchanged, as shown in Figure 9.16. Mitochondria stained with JC-1 were biopsied from 25+ NB4 cells and in all these cases mitochondrial membrane potential was observable, which indicated 100% mitochondria survival rate.

The survival rate of the remaining cell after mitochondria extraction was also examined. The cells were stained with Fluo-4 Calcium Imaging (F10489, Thermo Fisher Scientific) and calcium flux was detected before and after the extraction of mitochondria [6]. The size and shape of the micropipette tip, the amount of biopsied content from cell, and depth of penetration of the micropipette tip are the major factors that affect disruption of the cell during cell biopsy. With smaller micropipettes and less cell penetration, the mitochondria biopsy can be achieved in a minimally invasive way. A larger sized micropipette tip as well as larger aspirated volume of the cell can increase the chance of successful mitochondria biopsy, but meanwhile it increases cell disruption as well. Figure 9.17 shows a significant difference in calcium staining after performing the biopsy with an unbeveled, large micropipette tip (~4um). In contrast, Figure 9.18 indicates that the intensity of calcium flux remained

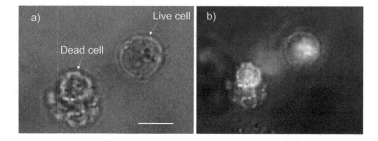

FIGURE 9.15 JC-1 staining of NB4 cells. (a) Bright field image. (b) Fluorescence image (GFP filter). (Source: A. Shakoor, M. Xie, T. Luo et al. / IEEE Transactions on Biomedical Engineering 66 (2019) 2210–2222, with permission.)

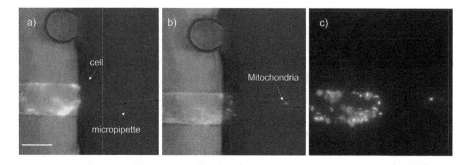

FIGURE 9.16 Functionality test of mitochondria after mitochondria biopsy. (a) Florence image before biopsy (GFP filter). (b) Florescence image after biopsy. (c) Florescence image after biopsy (RFP filter). (Source: A. Shakoor, M. Xie, T. Luo et al. / IEEE Transactions on Biomedical Engineering 66 (2019) 2210–2222, with permission.)

FIGURE 9.17　Cell (NB4) viability test with unbeveled and larger micropipette tip (~5um). (a) Under bright field imaging. (b) Before biopsy (Florescence image). (c) After biopsy (5sec). Scale bar is 15um. (Source: A. Shakoor, M. Xie, T. Luo et al. / IEEE Transactions on Biomedical Engineering 66 (2019) 2210–2222, with permission.)

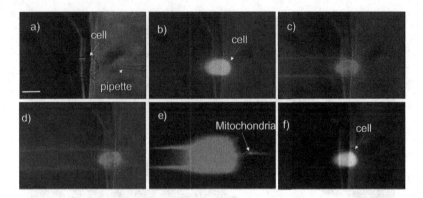

FIGURE 9.18　Cell (NB4) viability test. (a) Bright field imaging. (b) Florescence image before biopsy. (c) Mitochondria staining image (d) Image during biopsy. (e) Image after biopsy (arrowhead indicate aspirated mitochondria). (f) Calcium flux after biopsy (5sec). Scale bar is 15um. (Source: A. Shakoor, M. Xie, T. Luo et al. / IEEE Transactions on Biomedical Engineering 66 (2019) 2210–2222, with permission.)

approximately same when a beveled micropipette with a tip size of 1~1.5um (e.g. Figure 9.19) was used to perform mitochondria biopsy. We found that with our methodology, the mitochondria extraction rate and cell viability was inversely proportional. With the small micropipette tip and less aspiration of cellular content, less damage to the cell is seen. However, with these smaller micropipettes, the chance of aspiration of mitochondria into the micropipette tip is also reduced. Table 9.1 shows the extraction rate of mitochondria and nuclei of NB4 and HDF cells achieved with the automated organelle biopsy system. The organelle extraction rate is calculated from the ratio of the number of cells that are successfully biopsied to the total number of cells used in experiments. The mitochondria and nuclei biopsy for the NB4 cell could achieve an extraction rate of 62.4% and 68%, respectively. Similarly, the organelle extraction rate of HDF cells was 58% for the mitochondria and 72% for the nuclei. The mitochondria biopsy obtained a relatively low efficiency because in some cases, the detected mitochondria may be located near the back portion of the cell and the micropipette may be unable to perforate the membrane of the nuclei. Extraction rate of nuclei was affected by sliding of the micropipette tip over soft cell membrane

TABLE 9.1
Organelle Extraction Rate

Cell	Organelle	No. of Cells Used for Organelle Extraction	No. of Cells with Successful Organelle Extraction	Organelle Extraction rate (%)
NB4	Mitochondria	125	78	62.4
NB4	Nuclei	125	86	68
HDF	Mitochondria	50	29	58
HDF	Nuclei	50	36	72

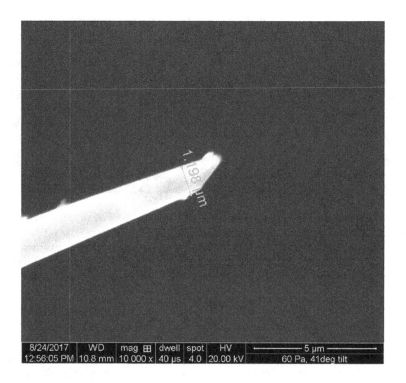

FIGURE 9.19   SEM Image of the beveled micropipette. (Source: A. Shakoor, M. Xie, T. Luo et al. / IEEE Transactions on Biomedical Engineering 66 (2019) 2210–2222, with permission.)

due to the relatively large size of the micropipette tip. Figures 9.20 and 9.21 summarize the results of the mitochondria extraction rate and the cell survival rate for different sizes of micropipette tip, respectively. The cell survival rate is calculated from the ratio of the number of cells that are successfully biopsied to the number of cells survived after the mitochondria biopsy. The negative pressure at the back of micropipette was calibrated while monitoring the successful aspiration of the mitochondria into the micropipette. A larger negative pressure corresponds to a large

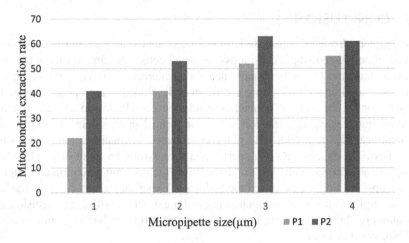

FIGURE 9.20 Mitochondria extraction rate with different sizes of the micropipette tip. (Source: A. Shakoor, M. Xie, T. Luo et al. / IEEE Transactions on Biomedical Engineering 66 (2019) 2210–2222, with permission.)

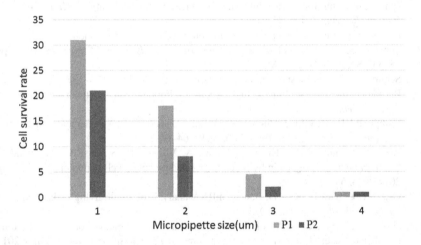

FIGURE 9.21 NB4 Cell survival rate with different sizes of the micropipette tip. (Source: A. Shakoor, M. Xie, T. Luo et al. / IEEE Transactions on Biomedical Engineering 66 (2019) 2210–2222, with permission.)

volume change inside micropipette, which consequently aspirates more cellular content during biopsy, and thus increases the mitochondria extraction rate. Based on experimental calibrations, two distinct aspiration pressures P1 and P2 determined by volume change per revolution were used in all experiments for examining the mitochondria biopsy and cell survival rates and results are summarized in Figures 9.20 and 9.21. P1 and P2 represent volume change by 20 and 45 degree revolution of injector, respectively.

9.5 CONCLUSION

This chapter introduces a case study of cell puncture technique contributing to realization of automated single cell biopsy for small cells with dimensions of less than 20 μm in diameter. The cells are patterned with a microfluidic device, and a template-matching-based image processing algorithm is developed to automatically measure the position of the desired organelles inside the cell. Followed by cell puncture manipulation, organelle extraction is then performed in an automatic way. A sliding nonlinear PID controller is developed to enhance the manipulation accuracy and robustness. The experiments were performed successfully on small suspended (NB4 ~18 μm) and adherent cells (HDF ~20 μm) in which the mitochondria and nucleus were biopsied using the proposed methodology. Experimental results show the successful implementation of automated organelle extraction with high precision and repeatability, thereby reducing human fatigue and error involved in complex single cell biopsy procedures.

REFERENCES

1. Liu J, Wen J, Zhang Z, Liu H, Sun Y. Voyage inside the cell: microsystems and nanoengineering for intracellular measurement and manipulation. *Microsystems & Nanoengineering* 2015; 1(1): 1–15.
2. Pertoft H, Laurent TC. Isopycnic separation of cells and cell organelles by centrifugation in modified colloidal silica gradients. In: *Methods of Cell Separation* 1977: pp. 25–65). Springer, Boston, MA.
3. Schubert C. The deepest differences. *Nature* 2011; 480(7375): 133–137.
4. Braude P, Pickering S, Flinter F, Ogilvie CM. Preimplantation genetic diagnosis. *Nature Reviews Genetics* 2002; 3(12): 941–953.
5. Campbell KH, McWhir J, Ritchie WA, Wilmut I. Sheep cloned by nuclear transfer from a cultured cell line. *Nature* 1996; 380(6569): 64–66.
6. Actis P, Maalouf MM, Kim HJ, Lohith A, Vilozny B, Seger RA, Pourmand N. Compartment genomics in living cells revealed by single-cell nanobiopsy. *ACS Nano* 2014; 8(1): 546–553.
7. Osada T, Uehara H, Kim H, Ikai A. mRNA analysis of single living cells. *Journal of nanobiotechnology* 2003; 1(1): 1–8.
8. Hayakawa T, Fukada S, Arai F. Fabrication of an on-chip nanorobot integrating functional nanomaterials for single-cell punctures. *IEEE Transactions on Robotics* 2013; 30(1): 59–67.
9. Singhal R, Orynbayeva Z, Kalyana Sundaram RV, Niu JJ, Bhattacharyya S, Vitol EA, … Gogotsi Y. Multifunctional carbon-nanotube cellular endoscopes. *Nature Nanotechnology* 2011; 6(1): 57–64.
10. Sakaki K, Esmaeilsabzali H, Massah S, Prefontaine GG, Dechev N, Burke RD, Park EJ. Localized, macromolecular transport for thin, adherent, single cells via an automated, single cell electroporation biomanipulator. *IEEE Transactions on Biomedical Engineering* 2013; 60(11): 3113–3123.
11. Dewan MAA, Ahmad MO, Swamy MNS. Tracking biological cells in time-lapse microscopy: An adaptive technique combining motion and topological features. *IEEE Transactions on Biomedical Engineering* 2011; 58(6): 1637–1647.
12. Xie M, Shakoor A, Shen Y, Mills JK, Sun D. Out-of-plane rotation control of biological cells with a robot-tweezers manipulation system for orientation-based cell surgery. *IEEE Transactions on Biomedical Engineering* 2018; 66(1): 199–207.

13. Wang Z, Feng C, Ang WT, Tan SYM, Latt WT. Autofocusing and polar body detection in automated cell manipulation. *IEEE Transactions on Biomedical Engineering* 2016; 64(5): 1099–1105.

14. Ladjal H, Hanus JL, Ferreira A. Micro-to-nano biomechanical modeling for assisted biological cell injection. *IEEE Transactions on Biomedical Engineering* 2013; 60(9): 2461–2471.

15. Sakaki K, Dechev N, Burke RD, Park EJ. Development of an autonomous biological cell manipulator with single-cell electroporation and visual servoing capabilities. *IEEE Transactions on Biomedical Engineering* 2009; 56(8): 2064–2074.

16. Zhang XP, Leung C, Lu Z, Esfandiari N, Casper RF, Sun Y. Controlled aspiration and positioning of biological cells in a micropipette. *IEEE Transactions on Biomedical Engineering* 2012; 59(4): 1032–1040.

17. Inayat S, Zhao Y, Cantrell DR, Dikin DA, Pinto LH, Troy JB. A novel way to go whole cell in patch-clamp experiments. *IEEE Transactions on Biomedical Engineering* 2010; 57(11): 2764–2770.

18. Haghighi R, Cheah CC. Optical manipulation of multiple groups of microobjects using robotic tweezers. *IEEE Transactions on Robotics* 2016; 32(2): 275–285.

19. Ogawa N, Oku H, Hashimoto K, Ishikawa M. Microrobotic visual control of motile cells using high-speed tracking system. *IEEE Transactions on Robotics* 2005; 21(4): 704–712.

20. Chen H, Sun D. Moving groups of microparticles into array with a robot–tweezers manipulation system. *IEEE Transactions on Robotics* 2012; 28(5): 1069–1080.

21. Liu J, Siragam V, Gong Z, Chen J, Fridman MD, Leung C, ... Sun Y. Robotic adherent cell injection for characterizing cell–cell communication. *IEEE Transactions on Biomedical Engineering* 2014; 62(1): 119–125.

22. Kim DH, Yun S, Kim B. Mechanical force response of single living cells using a microrobotic system. In: *IEEE International Conference on Robotics and Automation, 2004. Proceedings. ICRA'04. 2004* 2004; vol. 5: pp. 5013–5018). IEEE.

23. Huang HB, Sun D, Mills JK, Cheng SH. Robotic cell injection system with position and force control: toward automatic batch biomanipulation. *IEEE Transactions on Robotics*, 2009; 25(3): 727–737.

24. Wang WH, Liu XY, Sun Y. Contact detection in microrobotic manipulation. *The International Journal of Robotics Research* 2007; 26(8): 821–828.

25. Wang Z, Ang WT. Automatic dissection position selection for cleavage-stage embryo biopsy. *IEEE Transactions on Biomedical Engineering* 2015; 63(3): 563–570.

26. Wong CY, Mills JK. Cleavage-stage embryo rotation tracking and automated micropipette control: Towards automated single cell manipulation. In: *2016 IEEE/RSJ International Conference on Intelligent Robots and Systems (IROS)* 2016: pp. 2351–2356. IEEE.

27. Ichikawa A, Tanikawa T, Matsukawa K, Takahashi S, Ohba K. Automated cell-cutting for cell cloning. *SICE Journal of Control, Measurement, and System Integration* 2010; 3(2): 75–80.

28. Hagiwara M, Ichikawa A, Kawahara T, Arai F. High speed enucleation of oocyte using magnetically actuated microrobot on a chip. In: *2012 7th IEEE International Conference on Nano/Micro Engineered and Molecular Systems (NEMS)* 2012: pp. 364–367). IEEE.

29. Ichikawa A, Tanikawa T, Matsukawa K, Takahashi S, Ohba K. Fluorescent monitoring using microfluidics chip and development of syringe pump for automation of enucleation to automate cloning. In: *2009 IEEE International Conference on Robotics and Automation* 2009: pp. 2231–2236. IEEE.

30. Yu S, Nelson BJ. Microrobotic cell injection. In: *Proceedings 2001 ICRA. IEEE International Conference on Robotics and Automation (Cat. No. 01CH37164)* 2001; vol. 1, pp. 620–625. IEEE.

31. Sun Y, Nelson BJ. Biological cell injection using an autonomous microrobotic system. *The International Journal of Robotics Research* 2002; 21(10–11): 861–868.
32. Xie Y, Sun D, Liu C, Tse HY, Cheng SH. A force control approach to a robot-assisted cell microinjection system. *The International Journal of Robotics Research* 2010; 29(9): 1222–1232.
33. Viigipuu K, Kallio P. Microinjection of living adherent cells by using a semi-automatic microinjection system. *Alternatives to Laboratory Animals* 2004; 32(4): 417–423.
34. Wang W, Sun Y, Zhang M, Anderson R, Langille L, Chan W. A system for high-speed microinjection of adherent cells. *Review of Scientific Instruments* 2008; 79(10): 104302.
35. Kallio P, Ritala T, Lukkari M, Kuikka S. Injection guidance system for cellular microinjections. *The International Journal of Robotics Research* 2007; 26(11–12): 1303–1313.
36. Liu X, Kim K, Zhang Y, Sun Y. Nanonewton force sensing and control in microrobotic cell manipulation. *The International Journal of Robotics Research* 2009; 28(8): 1065–1076.
37. Chow YT, Chen S, Liu C, Liu C, Li L, Kong CWM, … Sun D. A high-throughput automated microinjection system for human cells with small size. *IEEE/ASME Transactions on Mechatronics* 2015; 21(2): 838–850.
38. Becattini G, Mattos LS, Caldwell DG. A fully automated system for adherent cells microinjection. *IEEE Journal of Biomedical and Health Informatics* 2013; 18(1): 83–93.
39. Chow YT, Chen S, Wang R, Liu C, Kong CW, Li RA, … Sun D. Single cell transfection through precise microinjection with quantitatively controlled injection volumes. *Scientific Reports* 2016; 6(1): 1–9.
40. Shakoor A, Luo T, Chen S, Xie M, Mills JK, Sun D. A high-precision robot-aided single-cell biopsy system. In: *2017 IEEE International Conference on Robotics and Automation (ICRA)*, 2017: pp. 5397–5402. IEEE.
41. Smiley ST, Reers M, Mottola-Hartshorn C, Lin M, Chen A, Smith TW, … Chen LB. Intracellular heterogeneity in mitochondrial membrane potentials revealed by a J-aggregate-forming lipophilic cation JC-1. *Proceedings of the National Academy of Sciences* 1991; 88(9): 3671–3675.
42. Kluza J, Gallego MA, Loyens A, Beauvillain JC, Sousa-Faro JMF, Cuevas C, … Bailly C. Cancer cell mitochondria are direct proapoptotic targets for the marine antitumor drug lamellarin D. *Cancer Research* 2006; 66(6): 3177–3187.
43. Pasini EM, Van den Ierssel D, Vial HJ, Kocken CH. A novel live-dead staining methodology to study malaria parasite viability. *Malaria Journal* 2013; 12(1): 1–10.

Index

Pages in *italics* refer figures and pages in **bold** refer tables.